The Culture of Tobacco

by George M. Odlum

with an introduction by Roger Chambers

This work contains material that was originally published in 1905.

This publication was created and published for the public benefit, utilizing public funding and is within the Public Domain.

This edition is reprinted for educational purposes and in accordance with all applicable Federal Laws.

Introduction Copyright 2018 by Roger Chambers

COVER CREDITS

Front Cover
Monte Cristo Cigar by Thermos (Own work)
[GFDL - http://www.gnu.org/copyleft/fdl.html]
or
[CC BY-SA 4.0-3.0-2.5-2.0-1.0 - https://creativecommons.org/licenses/by-sa/4.0-3.0-2.5- 2.0-1.0],
via Wikimedia Commons

Back Cover
Tobacco leaves drying by MRaccine (Own work)
[CC BY-SA 3.0 - https://creativecommons.org/licenses/by-sa/3.0]
or
[GFDL - http://www.gnu.org/copyleft/fdl.html],
via Wikimedia Commons

Research / Resources
Wikimedia Commons
www.Commons.Wikimedia.org

Many thanks to all the incredible photographers, artists,
researchers, biographers, historians, and archivists who share
their great work via the Wikipedia family.

PLEASE NOTE :
As with all reprinted books of this age that are intended to perfectly reproduce the original edition, considerable pains and effort had to be undertaken to correct fading and sometimes outright damage to existing proofs of this title. At times, this task can be quite monumental, requiring an almost total rebuilding of some pages from digital proofs of multiple copies. Despite this, imperfections still sometimes exist in the final proof and may detract slightly from the visual appearance of the text.

DISCLAIMER :
Due to the age of this book, some methods or practices may have been deemed unsafe or unacceptable in the interim years. In utilizing the information herein, you do so at your own risk. We republish antiquarian books without judgment or revisionism, solely for their historical and cultural importance, and for educational purposes.

Self Reliance Books

Get more historic titles on animal and stock breeding, gardening and old fashioned skills by visiting us at:

http://selfreliancebooks.blogspot.com/

Disclaimer

This book was written in an age when little was known about the ill effects of tobacco.

The material presented herein is intended to be strictly for educational purposes with the purpose of enlightening readers about the historical uses of tobacco. Publication of the material is neither an endorsement, nor a criticism of its contents. This book is presented as part of large series of educational material on the history and cultivation of tobacco.

As the reader, please consider it your duty to consult with a medical doctor before utilizing tobacco. It is also the reader's duty to become familiar with local, state, provincial and federal laws relating to the growing of tobacco.

As the author, publisher and retailer cannot control how the reader utilizes the historical information presented in the pages herein, they hereby disclaim any liability to any party for any loss, damage, disruption, death or other liability that may be incurred by the reader's misuse of this material.

introduction

Here at **Self-Reliance Books** we are dedicated to bringing you the best in *dusty-old-book-knowledge* to help you in your quest for self-sufficiency and independence.

We're so pleased to bring you this old title on the culture of Tobacco. These old books on agricultural and horticultural subjects are extremely popular. It should be said, though, that some of the information in a book like this is best looked at in the historical aspect, due to the obsolescence of some practices or methods.

This special edition of **The Culture of Tobacco** was written by George M. Odlum, and first published in 1905, making it well over one-hundred years old.

This old book features sections on *Tobacco Soils, Influence of Climate on Tobacco, Characteristics of a Good Tobacco, Cultivation, Insect Pests of Tobacco, Diseases of Tobacco*, and more.

A great old book and a must-read for all those interested in the historical aspect of the Tobacco Industry and the Tobacco plant as a crop, or those interested in the historical aspect of the Tobacco Industry.

~ *Roger Chambers*

State of Jefferson, March 2018

WASHINGTON, D.C., U.S.A.

SIR,

I have the honour to transmit to you herewith my report on tobacco culture. This report is mainly based on an investigation of the tobacco industry of America, which I have had the pleasure of making under your direction. I trust that the information embodied in this report may be of value to those who are endeavouring to develop the vast latent agricultural wealth of Rhodesia.

Respectfully submitted,

GEORGE M. ODLUM.

E. ROSS TOWNSEND, Esq.,
Secretary for Agriculture, Southern Rhodesia.

INTRODUCTION.

TO secure the information embodied in this report the writer has travelled some thirty-five thousand miles, has visited all the leading tobacco districts of America, has questioned the leading tobacco experts of the United States, and written nearly one thousand letters to verify facts, as well as studied all of the available tobacco literature.

The author is especially indebted to the Hon. James Wilson, Secretary of Agriculture of the United States, for courtesy extended in granting the freedom of his Department; and to Professor Milton Whitney, Chief of the Bureau of Soils and official head of the tobacco division of the United States Department of Agriculture, for information given and for his kindness in reporting on samples of Rhodesian soils and tobaccos. The other officials of the same Department have been uniformly obliging in giving information or assistance. The reports and bulletins of this Department, which contain the most exact and scientific agricultural information of any publications in the English language, have been fully utilized, as well as the Experiment Station reports of Maryland, Connecticut, North Carolina and Kentucky.

Mr. E. K. Vietor, of Richmond, Virginia, who is regarded as the leading authority on the export tobaccos, rendered assistance in many ways, and it was he who selected the leaves appearing in the coloured plates. Messrs. Patton and James, leaf dealers, of Darlington, South Carolina, also gave valuable and greatly appreciated assistance, as well as did Mr. G. M. Underhill, of Quincy, Florida. Taken all together one could not wish to meet a more agreeable body of men than those connected with the America tobacco industry.

No attempt has been made in this report to record all the details of tobacco culture in every section, for that would but lead to confusion on the reader's part. Only those features that seem worthy of adoption in a new country are described, except in those instances where they

INTRODUCTION.

are introduced for completeness' sake. The details of tobacco culture are capable of endless modifications, but the underlying principles of the production of fine tobacco are the same the world over, and because of this the effort has been to give the reasons for certain steps more prominence than the steps themselves. Where the nature of the tobacco plant is understood, and the principles involved in the production of fine tobacco are regarded, the methods may be modified to suit the changed conditions.

(The coloured plates in this volume were lithographed by the ANDREW B. GRAHAM COMPANY, *of Washington, D.C., U.S.A.)*

CONTENTS.

	PAGE.
Tobacco	1
Tobacco soils	2
Influence of climate on tobacco	4
The tobacco plant	6
Varieties	6
Classification of tobaccos	24
Characteristics of a good tobacco	31
The tobaccos called for by different countries	33
The seed bed	38
Preparation of the land, and planting	42
Cultivation	45
Topping	46
Priming	47
Suckering	47
Ripening	48
Harvesting	50
Transportation from the field	56
The growth and selection of tobacco seed	60
Insect pests of tobacco	63
Cut worms	65
Horn caterpillars	65
Bud caterpillars	70
Leaf miner or "split worm"	70
Tobacco flea beetle	70
A beetle injurious to stored tobacco	71
Other tobacco insects	71
Diseases of tobacco	71
Mosaic disease	71
Frog eye or leaf spot	72
Rust or blight	72
Diseases of tobacco while curing	72
Pole burn, pole sweat, or house burn	72
Stem rot	73
White veins	73
Moulds and rots in cured tobacco	73
"Saltpetre"	74

CONTENTS.

	PAGE.
Damage by hail and wind	74
Curing: General remarks	74
Sun curing	76
Fire curing	77
Flue curing	78
Air curing	82
Cigar leaf curing	82
Fermentation of cigar leaf	85
Bulk fermentation	85
Ageing	93
Some of the chemistry of curing and fermentation	93
Buildings	96
Curing barns	96
Stripping, grading or packing houses	110
Fermentation houses	110
Packing or prizing of all tobaccos except cigar leaf	110
Marketing tobacco in America	113
Loose sales	113
The package and sampling system	120
Sales of cigar leaf	122
Leaf dealers or middlemen	122
Re-ordering and stemming tobacco	122
Stemming or stripping	130
Some of the chemistry of the tobacco plant and its relation to the fertilizers used	130
Water in its relation to the tobacco crops	138
Production of tobacco under shade	142
Tobacco as an insecticide	150
Perique tobacco culture	150
Tobacco culture in Sumatra	155
Cost of production, profits, wages and yields	160
A few concluding words	163
Appendix and statistical tables	167

INDEX TO COLOURED PLATES OF TOBACCO LEAVES.

Burley	to face page	viii
Connecticut Havana Seed Leaf	,, ,, ,,	14
Bright Lemon, for Cigarettes	,, ,, ,,	28
Mahogany Leaf, for Plug Wrappers	,, ,, ,,	32
Dark Vuelta Abajo	,, ,, ,,	74
Cuban Cigar Wrapper	,, ,, ,,	82
Dark English Stemming	,, ,, ,,	130
Light Sumatra Cigar Wrapper	,, ,, ,,	154

THE CULTURE OF TOBACCO.

SINCE the day when Columbus observed the natives of his newly-discovered Western Land whiffing its fragrant leaves, tobacco has been making the conquest of the world, until to-day but few peoples, civilized or savage, have not become the devotees of the pipe, or the consumers of the weed in some form. The peace pipe of the red man has become the peace pipe of the white man, and the black man, and the yellow man.

While this seductively aromatic and mildly narcotic plant has been charming the senses of man, it has also been producing the wealth of its culturists, and increasing the revenues of nations. It has assisted at the birth of new lands, and aided in the maintenance of old ones.

The plant thrives in nearly every portion of the world, yet the nations that consume it most largely have been strangely slow in adopting its culture, or in learning the intricacies of its production. America, its natal continent, with its virgin soil, its cheap lands, and its people trained in its culture, remains the mother of the industry to-day, and from the shores of this continent sail fleets, to carry this solace to far-off peoples, and to bring back wealth for its producers. But the soils of America will soon be no longer virgin, their fertility will have been carried away to other countries, the value of the lands will be higher, and the cost of production will be increased. Then strange will it be if the culture of tobacco be not wrenched away from its mother land, and the pipes of the world supplied from fields newly discovered, and now yielding for the first time to the hand of man.

The primitive methods of culture employed by the red man and the early colonists have yielded to methods highly complex, and better adapted to the production of an article suited to the educated tastes of modern peoples. Coarse leaves dried on the bushes in the streets of Jamestown would not receive the approval of a twentieth-century Raleigh.

Science has recently been summoned to the aid of experience, with the result that this marvellous plant, with its relation to the soil, climate, culture and palate of its ardent consumers, has at last divulged many of its long and tenaciously kept secrets.

To-day, as never before, it invites students of nature, friends of commerce, and lovers of pleasure to examine its ways, test its merits, and afford it an opportunity to prove itself the friend of man.

TOBACCO SOILS.

The tobacco plant readily adapts itself to a great variety of soils. It can, in fact, be produced on any soil where other agricultural crops will thrive; and yet there is no other plant so easily affected by the chemical and mechanical conditions of the soil, for, while the tobacco plant will adapt itself to diverse conditions of

soil and of climate, still each distinct type requires certain conditions to give to it those qualities of colour, texture, and aroma for which it is prized.

The relation of the physical condition of the soil to the texture and quality of the leaf has been so well established that soil experts are now able to go into a non-tobacco producing section of the country, and, by a mechanical analysis of the soil and a determination of its moisture-holding capacity, state very nearly the adaptability of the section to any particular class or type of tobacco. The chemical condition of the soil has largely to do with the burning qualities of the leaf as well as with the rapidity of growth of the developing plant.

The colour of a soil is largely indicative of its mechanical and, to some extent, its chemical condition. Light coloured soils generally produce bright coloured tobaccos, and dark soils dark coloured tobaccos. Soils containing a large proportion of clay, or which have a large moisture-holding capacity, produce heavy tobacco which cures to a dark brown or red; while soils consisting largely of sand produce tobacco that cures out a yellow or bright colour. Often the clay subsoil will be exposed in some portion of a yellow field, and on this exposed portion the tobacco produced will be of a darker colour. Very rich soils that will produce a large leaf will usually produce a tobacco of poor quality.

An attempt to produce a tobacco on a soil not suited to the type of tobacco planted will, in most cases, meet with failure, for the tobacco produced is unfit to place in the same class as the parent plant, and at the same time it is not likely to grade with any other established type, and as a result is unclassed, and sells as nondescript. It is only the exceptional case where a new type is thus established worthy to create a market on its own merits. Seed of the dark export varieties of tobaccos, if planted on the light soils adapted to the yellow tobaccos, will not be bright enough to class with those tobaccos, nor dark enough to class with the exports, and furthermore, the yield will not be as heavy as it would be if the varieties had been planted on their own heavy soil.

The White Burley tobacco of Kentucky is grown on a well-drained deep red soil, the surface of which is of a light loamy character, and not likely to clod when properly worked. These lands are fairly rich in lime, and produce almost ideal crops of maize, wheat, hemp and grass. The subsoil contains about 30 per cent. of clay, and has a moisture content of about 20 per cent.

The Bright tobacco lands of Virginia and North and South Carolina consist of sand of varying density. This soil contains not more than 8 to 10 per cent. of clay, and is usually underlaid with a red or yellow clay subsoil, at a depth of about a foot. The deeper the sand, the brighter the tobacco produced, and the nearer the surface that the subsoil comes, the more inclined the tobacco is to darken and be mahogany in colour, until where the subsoil is completely exposed, the tobacco produced is altogether dark. However, where the sand is very deep, there is not the same surety of a sufficient supply of moisture during the growing season, and for this reason it is preferred that the subsoil be within 18 inches of the surface.

The heavy dark export types of tobacco produced in Tennessee and Kentucky are grown on a rich well-drained soil, containing about 50 per cent. of silt, 23 per cent. of clay, and have an average moisture content of 15 per cent. These lands are underlaid with a red clay subsoil, and are fairly well supplied with lime. They produce heavy crops, but deteriorate rapidly, unless the land is kept up to its original condition, by the addition of fertilizers or by methods of cultivation.

The cigar leaf lands of Connecticut consist of light, alluvial, sandy soils containing a small percentage of clay, and, as a rule, the less the percentage of clay the greater the percentage of fine cigar wrappers. A few years ago, when a darker coloured cigar was the fashion, this class of land was not used, but a much heavier soil, with a moisture-holding average of from 25 to 27 per cent., was cultivated, whereas the soil now used for the production of light wrappers has a moisture content of but 7 per cent. Where cigar filler leaf is grown, and the colour, texture, and elasticity are not points so important as with the wrappers, a rich clay is used, and a very heavy crop of leaf secured. The Sumatra seed tobacco is produced in this section on the lighter soils.

In the cigar leaf section of Ohio no attempt is made to produce anything but a filler tobacco. The soil that produces the highest quality of leaf is a thin silty soil resting on a red clay or a silty subsoil. The tobacco lands of this section contain, on an average, 25 per cent. of moisture, which would be too much for the production of a wrapper leaf. In this region any attempt to grow tobacco on the rich black lands results in a coarse, heavy, badly flavoured leaf.

In Pennsylvania there are two types of tobacco soils, the light alluvial soils similar to those in Connecticut, on which a fair wrapper leaf is grown, and the heavy clay limestone, on which heavy crops of cigar filler are produced. The latter soil contains an average of 30 per cent. of clay and 20 per cent. of moisture.

The Wisconsin leaf is grown on a well-drained dark rich loam, underlaid by a heavy silt or clay. This leaf is chiefly used as cigar binders.

The cigar leaf soil of Western Florida is a light sand loam underlaid by a red clay subsoil, and closely resembles the yellow tobacco lands. This soil very successfully produces both the Cuban and the Sumatra tobacco. Sumatra tobacco is grown for the production of high-grade wrappers, and these soils contain from 8 to 10 per cent. of moisture, an amount very similar to that in good wrapper-producing soil elsewhere. Both the Cuban and the Sumatra tobaccos are also being grown further south in Florida on a somewhat coarser sand, which contains about the same percentage of moisture.

Perique tobacco is grown in Louisiana on a deep, gray, fertile loam. This soil is well drained, friable, and retentive of a medium percentage of moisture. It not only forces a rapid growth of the plant, but helps to give to it that gumminess necessary to develop the characteristic Perique aroma.

It will be noticed that the texture of the tobacco grown, and the purpose for which it is adapted, is dependent on the moisture-holding average of the soil. A soil with a moisture content of over 12 per cent. is not adapted to the production of wrappers. However, the texture of the soil does not account for differences in combustibility and in aroma; for these we must look for differences in the chemical constituents of the soil, and variations in climate.

CLIMATE.

Few plants are so susceptible to climate as is tobacco. Climate largely influences the quality and aroma in the same way that soil influences the texture. In a warm climate the tendency of the leaf is to be gummy, resinous and aromatic. In a cooler climate the leaf will become larger, thinner, and almost without aroma. While the tendency is for the leaf to become thick in warm climates, this tendency may be overcome by other conditions as excessive rainfall. This is the condition in Sumatra, where the leaf is famous for its fineness of texture. Tobacco grown in regions of excessive rainfall is washed out and devoid of fine aroma. This is understood in Cuba, where the tobacco is not planted until the cessation of the heavy rains. Cold rainy weather increases the acidity of the leaf, and this may have a detrimental effect on the curing and fermentation processes by preventing the action of the oxydizing enzymns. Excessively dry weather also prevents the formation of enzymns, and tobacco grown in dry climates is not likely to develop a fine aroma. A moderate rainfall with warm weather is perhaps the best condition for the production of tobacco. Dry weather during the ripening period is favourable to the preservation of those products that later create the aroma of cured tobacco. This is particularly true if the dry weather be accompanied with heavy dews. The dews incite the leaf to the formation of gums and resins. Hot dry weather causes a greater thickening of the leaves than does moist weather, and leaves grown in the shade are thinner than those grown in the direct sunlight.

Proximity to the sea has a great influence on the quality of the product. Tobacco grown near the sea is poor in combustibility. This is supposed to be due to the action of chlorine in the salt of the sea air. At thirty miles from the coast this influence may be said to have ceased, and in Sumatra good tobacco is grown within ten miles of the coast. The Italian Government will not permit the cultivation of tobacco on land with an elevation less than one hundred and ten feet. This is probably due to the fact that the low lands are near salt water. In dry climates proximity to a body of water, particularly to fresh

water, may be an advantage in that it will increase the humidity of the air.

In Sumatra better tobacco is produced on low well-drained lands some distance from the coast, than is produced further back on the mountain slopes. In this instance, texture rather than aroma is the feature sought for, and it may be owing to a difference in the quality of the soil rather than to difference in elevation. The finest tobacco in the world from the standpoint of aroma is grown in the mountain valleys of Western Cuba. The altitude of these valleys is not at all excessive. The supposition is that high elevations are not likely to produce fine tobacco, for heat is an essential to the development of aroma, and the high altitudes are generally cool.

In general, tropical climates will produce aromatic tobaccos, which are the best for cigar fillers, and the cooler portions of the temperate climates will produce thin leaves with but little aroma which are adapted for cigar wrappers. The bulk of the pipe and chewing tobaccos of the world are produced in the warmer portions of the temperate zone. The plant can be grown in any place with two months of weather without frost, but the aroma will depend largely on the temperature and humidity conditions of the district.

THE TOBACCO PLANT.

The tobacco of commerce is produced from several different species of the *Genus Nicotiana*, of which there are some fifty species. *N. tabacum* is the species commonly cultivated in North America.

N. rustica originally came from South or Central America, where it is grown to a certain extent to-day. It is cultivated also in Germany, Hungary, and Russia, and probably furnishes the Turkish and Latakia tobaccos.

N. repanda furnishes the Yana tobacco of Cuba, which is cultivated to a very limited extent, the Havana tobacco being a variety of the *N. tabacum*. The Shiraz tobacco of Persia is a product of the species *N. Persica*. Many different varieties of the more important species have been developed, and it may be that some varieties are the result of a cross of two or even more of the species.

Tobacco is a member of the family *Solanaceæ*, to which also belongs the tomato, potato, pepper, egg plant, petunia, Cape gooseberry, the datura, and many other common plants, a fact it is well to remember in the cultivation of tobacco or in the combating of its enemies; for the insect pests, or diseases, of one member of the family are often the enemies of other members.

SELECTION OF VARIETIES.

Before entering into the production of tobacco the planter should give due thought to the selection of the varieties to be grown. Questions of soil and climate, as well as of market, should be taken into consideration, and the variety selected that will produce the type desired, as well as the largest proportion of the best grades of that type. By years of careful selection, varieties adapted to the different conditions of soil and climate have been developed, and out of these a small list should be chosen for trial of such as have proved to be the best under the same conditions now confronting the planter. Experience will determine the ones to be retained as standard croppers.

Tobacco is very susceptible to changes of locality. If a dozen different varieties be secured from different portions of the earth, and all planted under the same conditions, they all will have a tendency to become alike and to adapt themselves to the new locality. This does not mean, however, that any seed so long as it is tobacco will do to plant, for this adaptation requires years. It is far better to select varieties that will necessitate but little change to become established.

FROM "TOBACCO LEAF, ITS CULTURE, CURE AND MANUFACTURE,"
ORANGE JUDD COMPANY, NEW YORK.

In some instances the tobacco will retain most of its finer characteristics for but a few years. In this case it is advisable to frequently import fresh seed from the place where it reaches its highest excellence. Where Cuban tobacco is grown in the United States fresh seed is regularly imported from the Vuelta Abajo. The imported seed is not used directly for the planting of the main crop, but is sown for the production of a seed crop from which the main crop of the following year is grown. If produced year after year from local-grown seed the Cuban plant will tend to lose its fine aroma and become like the seed-leaf varieties.

BRIGHT TOBACCO IN SOUTH CAROLINA; TWO OR THREE SEED PLANTS IN THE FOREGROUND.

There are perhaps a hundred varieties of tobacco regularly grown in the United States, many of which are very similar. Slight changes in soil and climate produce many variations. The following descriptive list includes the most common varieties:—

ADOCK.—Wide space between leaves; ripens uniformly from top to bottom; used for wrappers and fillers for plug; excellent fine smokers; grown in North Carolina.

FROM "TOBACCO LEAF, ITS CULTURE, CURE AND MANUFACTURE,"
ORANGE JUDD COMPANY, NEW YORK.

BADEN.—Short leaves, light; inclined to be chaffy; cures a fine yellow, but liable to green spots; used for plug wrappers and fillers; smokers; grown in Maryland.

BALTIMORE CUBA.—Long leaf; good body; fine, silky texture; tough; yields well; sweats a uniform colour; disseminated by the U.S. Agricultural Department; grown in Ohio (Miami Valley).

BAY.—Large heavy leaf; red spangled and yellow when cured; used for manufacturing and shipping; grown in Maryland.

FIELD OF WHITE BURLEY, CHARTER DISTRICT, SOUTHERN RHODESIA.

BEAT-ALL (same as Williams).—Large spreading leaf; fine fibre; rich, dark and gummy; export to Great Britain and Germany; well cured; makes fine wrappers; grown in Tennessee and Virginia.

BELKNAP.—Sub-variety of Connecticut seed-leaf; used same as Connecticut seed-leaf; grown in Connecticut, Massachusetts and New York.

FROM "TOBACCO LEAF, ITS CULTURE, CURE AND MANUFACTURE,"
ORANGE JUDD COMPANY, NEW YORK.

BONANYA.—A white Burley cross on yellow Orinoco; said to possess the qualities of both parents; very hardy; popular for yellow and mahogany leaf.

BULL-FACE.—Sub-variety of the Pryor; large, heavy leaf, oval shaped; tough; small stems and fibres; a luxuriant grower; used for heavy shipping; makes good wrappers for plug; grown in Virginia, North Carolina, and Tennessee.

BULLION.—A white Burley cross on Hester; a broad-leaved stately plant, well formed and fine fibred; has fine texture and great absorptive capacity; for yellow and mahogany leaf.

BURLEY.—*Red.*—Thin leaf narrowing toward the tip from centre; used for cutting tobacco; grown in Kentucky, Virginia and Ohio.

BURLEY.—*White.*—Long, narrow leaf; white in appearance while growing; grows flat with points of leaves hanging down; used for fancy wrappers and for cutting purposes; grown in Ohio, Kentucky, Virginia, Maryland, Missouri and Indiana.

FROM "TOBACCO LEAF, ITS CULTURE, CURE AND MANUFACTURE," ORANGE JUDD COMPANY, NEW YORK.

CLARDY.—Large, smooth, heavy leaf, extremely broad; stalks long; a hybrid; used for common plug; exported for Swiss wrappers; grown in Kentucky and Tennessee.

CONNECTICUT SEED-LEAF.—Broad leaf; strong, thin, elastic, silky; small fibres; used for cigar wrappers, lower grades for binders and fillers; grown in Connecticut, New Hampshire, New York, Pennsylvania, Ohio, Wisconsin, Minnesota, also in Indiana, Illinois and Florida.

CONNECTICUT BROAD LEAF.—Modification of above; leaves broader in proportion to length; fibres more at right angles to

THE CULTURE OF TOBACCO. 13

CUBAN TOBACCO IN SOUTHERN FLORIDA.

mid-rib; used as above; grown in Connecticut, New York and Wisconsin.

CUBA.—Small leaf, grown from imported seed; retains much of the aroma of Cuba-grown tobacco; used for cigar wrappers, fillers and binders; grown in Pennsylvania, New York, Ohio, Wisconsin, Florida and Louisiana.

CUNNINGHAM.—Short, broad leaf; thick and stalky growth; fillers and smokers; grown in North Carolina.

DUCK ISLAND.—Broad leaf; fine appearance; full grower; originated from Havana seed; used for cigar work; grown in New York and Pennsylvania.

FLANNAGAN.—Similar to little Orinoco, but broader leaf; finer fibre; silky and tough; used for fancy wrappers and plug fillers; grown in Virginia.

FLORIDA LEAF.—Fine texture, silky and elastic; becomes spotted with white when ripening; used for cigar wrappers, binders and fillers; grown in Florida.

FREDERICK.—Akin to White Stem; rough leaf; heavy and rich; stands up well; used mainly for export to Europe; grown in Virginia and Tennessee.

GLESSNER.—Large handsome leaf; fine texture; soft and elastic; used for cigar wrappers and fillers; smokers; grown in Pennsylvania, New York and Wisconsin.

GOOCH.—Broad, round leaf; leaves thick on stalk; yellow on hill when ripe; cures easily; used for fancy wrappers and smokers; grown in Virginia and North Carolina.

GOURD LEAF.—Broad, short, fine and silky leaf; used for plug wrappers and fillers, smokers; grown in Virginia.

GOVERNOR JONES.—Long, narrow leaf of good body; used for plug wrappers and fillers, and for common suckers; grown in Kentucky.

HAVANA SEED.—Very thin, fine leaf; fine texture; delicate flavour; used for cigar wrappers in Connecticut, Massachusetts, Pennsylvania and Wisconsin.

HESTER.—A broad shouldered, heart-shaped leaf, fine fibred, silky; cures very bright; for plug fillers and wrappers, yellow and mahogany tobacco.

HICKORY LEAF.—Fine fibre and texture; cures up very bright; used for plug work, smokers and shipping; grown in West Virginia.

HONDURAS.—Vigorous plant, one of the best for mahogany leaf.

JOHNSON GREEN.—Said to be a cross of Orinoco and White Stem: large, heavy leaf, strong flavour; used for strips and shipping leaf; grown in Virginia.

KENTUCKY YELLOW.—A prolific variety for the production of yellow tobacco.

KITE FOOT.—Rather short, wide leaf, thin; apt to cure a greenish colour unless fully ripe; used for very common cigars, culture decreasing; grown in Indiana.

LITTLE DUTCH.—Narrow leaf, small and short; in flavour resembling Yara tobacco; used for binders and fillers for cigars; grown in Ohio (Miami Valley).

LONG GREEN.—Coarse and heavy; vigorous grower; used for shipping leaf; grown in Virginia.

PENNSYLVANIA SEED-LEAF.

LANCASTER BROAD LEAF.—Upright grower; delicate silky fibre; used for cigar wrappers, binders and fillers; smokers; grown in Pennsylvania and Wisconsin.

FROM "TOBACCO LEAF, ITS CULTURE, CURE AND MANUFACTURE,"
ORANGE JUDD COMPANY, NEW YORK.

LOVELADY.—Long, dark, narrow leaf; very heavy; used for export; grown for African shippers; grown in Virginia, Tennessee and Indiana.

MANN.—Leaf of good body; heavy and gummy; used for plug wrappers and fillers; export; grown in North Carolina.

ORINOCO.—Short, broad leaf; upright growth and open habit; light coloured; much ruffled; used for plug wrappers and fillers; for strips and for export leaf; grown in Virginia and Missouri.

FROM "TOBACCO LEAF, ITS CULTURE, CURE AND MANUFACTURE,"
ORANGE JUDD COMPANY, NEW YORK.

ORINOCO (Big).—Short, broad leaf; doubtless the same as last named; used for sweet plug wrappers and fillers; export; grown in Virginia, Missouri, North Carolina, Tennessee and West Virginia.

ORINOCO (Little).—Long, narrow, tapering leaf; fine texture; stands up well; used principally for plug work and smokers; sweetest variety grown; grown in Virginia, North Carolina, Tennessee, West Virginia and Missouri.

PENNSYLVANIA SEED-LEAF.—Same as Connecticut seed-leaf; used same as Connecticut seed-leaf.

PERIQUE.—Medium-sized leaf; fine fibre; small stem; tough, gummy and glossy; used for smoking; cigars and cigarettes; for mixing with other kinds; grown in Louisiana.

PITTSYLVANIA YELLOW.—Medium size; leaves elongated, good distance apart; fine texture; small tough stem; used for fine wrappers and fillers; good export variety; grown in West Virginia.

PRYOR (Blue).—Large, fine leaf, long and well proportioned; good colour; slightly ruffled; used for cigar and plug fillers; grown in Virginia, North Carolina, Kentucky, Tennessee, Missouri and Indiana.

PRYOR (Yellow).—Heavy, wide leaf; fine texture; bright fine colour; tough; weighs well; used for cigar and plug wrappers and

FROM "TOBACCO LEAF, ITS CULTURE, CURE AND MANUFACTURE,"
ORANGE JUDD COMPANY, NEW YORK.

fillers; stemmers for export; grown in Virginia, North Carolina, Kentucky, Tennessee, Missouri and Indiana.

PRYOR.—White (or Medley Pryor).—Very broad leaf; soft and silky texture and tough fibre; a beautiful grower used for plug wrappers and fillers; grown in Virginia.

SHOESTRING.—Heavy leaf; rather narrow; long and large stem; used for dark navy plug; good stripping leaf; grown in Tennessee, Kentucky, Missouri and Virginia.

ZIMMER SPANISH CIGAR LEAF, OHIO.

20   THE CULTURE OF TOBACCO.

CUBAN TOBACCO ALONG THE LINE OF THE SOUTHERN PACIFIC IN TEXAS.

FIELD OF HEAVY SHIPPING TOBACCO; BARN IN THE DISTANCE; CLARKSVILLE, TENNESSEE.

22   THE CULTURE OF TOBACCO.

HEAVY SHIPPING TOBACCO.

STOCK-STEM.—Large, long leaf; heavy weigher; no ruffles; used for heavy dark fillers; shipping leaf; grown in Tennessee.

SUMATRA.—Imported from the Island of Sumatra. Grown in Florida and Connecticut; leaf thin, of fine texture, veins small; popular as cigar wrappers. Plant will grow seven or eight feet high under favourable conditions.

THICKSET.—Leaf long, pointed, narrow; coarse fibre; very short stalk; coarse and heavy; used for common plug work and shipping; grown in Kentucky, Missouri, Maryland, West Virginia, Tennessee and Eastern Ohio.

TWIST-BUD.—Heavy, large leaf; screw-shaped terminal stem; export mainly, also for plug fillers; grown in Kentucky, Missouri, and Maryland.

VALLANDIGHAM.—Large, pointed, smooth leaf; used for cigar wrappers and fillers, smokers; grown in Wisconsin.

VUELTA ABAJO.—The highest quality of cigar filler in the world; home is in the valleys of the western end of Cuba; small plants; grown in some portions of the United States, but loses its fine qualities after a few years.

WARNE.—Very popular new variety, for the production of yellow tobacco.

WHITE STEM.—Leaf long, slender, drooping; tough and fibrous; largest leaf grown; used for plug wrappers, strips, and shipping leaf; grown in Virginia.

WILLIAMS.—Same as "Beat-All"; grown in Tennessee for twenty-five years as "Williams"; British and German export; grown in Tennessee.

WILSON'S HYBRID.—Said to be an improved Havana; erect habit; easy of cultivation; used for cigar wrappers, binders, and fillers; grown in New York.

YELLOW MAMMOTH.—Very large leaf; rapid grower; yields largely; stemmed for export for Swiss wrappers; grown in Tennessee.

ZIMMER'S SPANISH.—Uniform dark colour; medium size leaf; high flavour; highly prized for cigar fillers; grown in Ohio.

It is impossible to state which of these varieties would be the best for another country, that being a matter that can be settled only by experiments. But if the soil and moisture conditions be known, it is easy to select a list in which the variety best adapted to the new conditions may be found. In the following list the very best varieties grown in America are grouped with reference to their uses and to the soils for which they are best adapted.

For the production of yellow and mahogany cutters, plug wrappers and fillers, by the method of flue-curing, and for growth on sandy soils—Conqueror, Warne, White Stem Orinoco, Yellow Orinoco, Long Leaf Gooch, Hester, Yellow Pryor, Eastern Pride, Bonanza, Gold Finder, Hyco, Granville, County Yellow.

For rich limestone soils and the production of pipe, cigarette, and chewing tobacco—White Burley.

For heavier soils and the production of sweet chewing tobacco—Sweet Orinoco and Flannagan. All the bright tobaccos in the first list will also produce a certain amount of chewing tobacco, and particularly plug wrappers.

For heavy, rich soils, and the production of a red, reddish-brown or dark tobacco for smoking and chewing, and particularly intended for the European and West African Market—Yellow Mammoth, Tennessee Red, Clardy, Kentucky Yellow, Medley Pryor, Blue Pryor. All of these produce a great weight of tobacco if grown on rich soil.

For the production of cigars, plant the following varieties:—(If grown on soils with a low water content they will produce a large percentage of cigar wrappers, and if grown on soils with a greater water content they will largely produce fillers.) Sumatra, for wrappers only; Vuelta Abajo, for high-class cigars; choice Havana and Cuban, for good cigars; Havana seed leaf, Connecticut seed leaf, Zimmer Spanish, Pennsylvania seed leaf, Pumpelly, and Brazilian for ordinary cigars or for blending with better grades.

This point should always be remembered—the darker and heavier the soil, the darker and heavier the leaf, no matter what variety be planted.

Classification of Tobaccos.

The classification of tobaccos according to the variety grown is of little value to the trade, because of the endless modifications produced by differences of soil and climate. The same variety grown on the same field for two different seasons may produce leaf that is adapted for entirely different purposes. The first season may be such that the leaf will be dry and thin, and only adapted for cigarettes, while the second season may grow a leaf that will be ideal for plug wrappers. The same plant will also produce several grades of leaf that will belong to more than one class; the lower leaves may be adapted for pipe smoking, the next for cigarettes, the middle leaves for plug wrappers, and the tips for a low grade of pipe tobacco, or if well ripened, for plug fillers. For this reason a classification quite distinct from the variety classification is adopted.

By a class is meant the purpose for which the tobacco is to be used, for cigars, for chewing, for cigarettes, or for export. A type is based on the combination of certain qualities and properties in the leaf, as colour, strength, elasticity, flavour, body, etc., or on certain characteristics produced by methods of curing, as air cured, sun cured, or flue cured. One type may often be placed in more than one class, as is the case with the yellow tobaccos which fall into both the smoking and chewing classes. One district may produce several types, and one or more of these types may be identical with certain types produced in other districts. However, the various types are usually confined to certain districts where the conditions are favourable for the development of qualities that give the leaf a distinct characteristic.

A grade is a sub-division of a type based on the different degrees of quality, texture, size, aroma, etc. These sub-divisions are nearly endless, for a crop may be divided into say five groups on a division based on quality, then each of these groups may be divided into say three sub-groups on a second selection based on colour, and each sub-group may again be divided into a dozen grades according to length. Sumatra tobacco from the same farm may be divided into seventy-two grades.

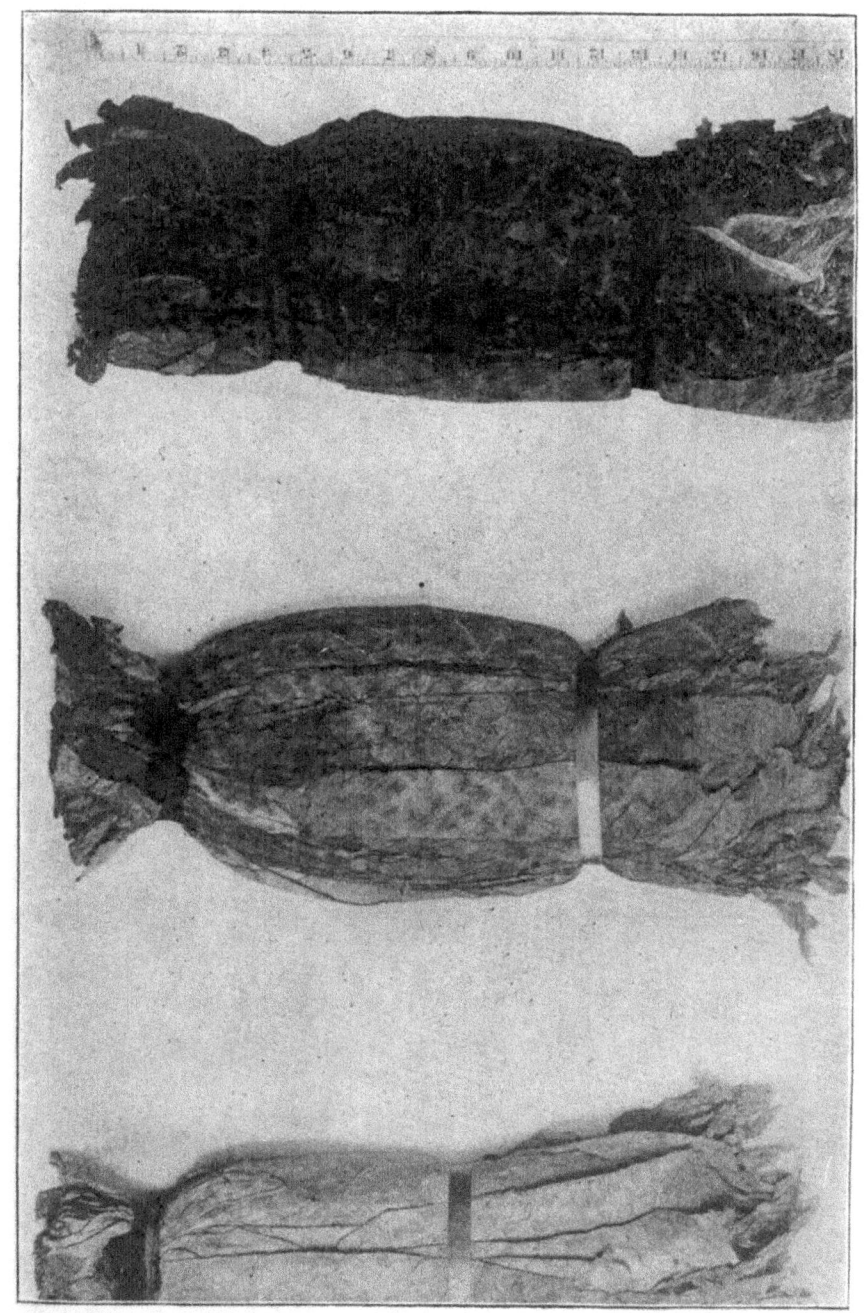

DARK MAHOGANY, LIGHT MAHOGANY AND LEMON WRAPPERS (U.S. DEPT. OF AG.).

26 THE CULTURE OF TOBACCO.

MARYLAND FINE LUG (U.S. DEPT. OF AG.).

To become expert in the classification and grading of tobacco requires lifelong experience. All that the farmer can attempt to do is to place all leaves of a certain size, colour, and quality together, and let the buyer classify them as he wishes. This proper assortment of the leaf is one of the most important things in the whole tobacco business. A few leaves placed in the grade above where they belong will largely destroy the selling value of the whole grade. In case of doubt, always place the leaf in the next grade below. Many farmers do not receive more than from half to three-quarters the value of their crop, for the reason that they have neglected to properly classify and grade their tobacco. Hundreds of shrewd leaf dealers have made their fortunes by buying up this poorly graded tobacco and regrading it. Once the crop has been grown, the planter should endeavour to secure its full value by placing it in condition to suit the buyer.

The following is a list of the types of tobacco produced in America; this list was prepared by the U.S. Department of Agriculture.

Types of Tobacco Grown in The United States.

Cigar Types.

Zimmer Spanish; grown in Ohio.
    Fillers, 9-18 inches.
Little Dutch; grown in Ohio.
    Fillers, 12-21 inches.
Ohio seed-leaf; grown in Ohio.
    Wrappers and binders, 16-26 inches.
Wisconsin Binders; grown in Wisconsin.
    Binders, 14-24 inches.
Pennsylvania Broadleaf; grown in Pennsylvania.
    Binders and Fillers, 14-26 inches.
Connecticut Broadleaf; grown in Connecticut.
    Wrappers, 14-26 inches.
Connecticut Havana; grown in Connecticut Valley.
    Wrappers, 14-26 inches.
York State Havana; grown in New York.
    Wrappers, 14-26 inches.
Connecticut-grown Sumatra.
    Wrappers, 12-18 inches.
Florida-grown Sumatra.
    Wrappers, 10-14 inches.
Cuban; grown in Florida, Texas, California (southern); also being experimented with in Ohio, Pennsylvania, Alabama, and South Carolina by U.S. Department of Agriculture.
    Wrappers, (very few), 12-16 inches.
    Fillers, 8-14 inches.

28 THE CULTURE OF TOBACCO.

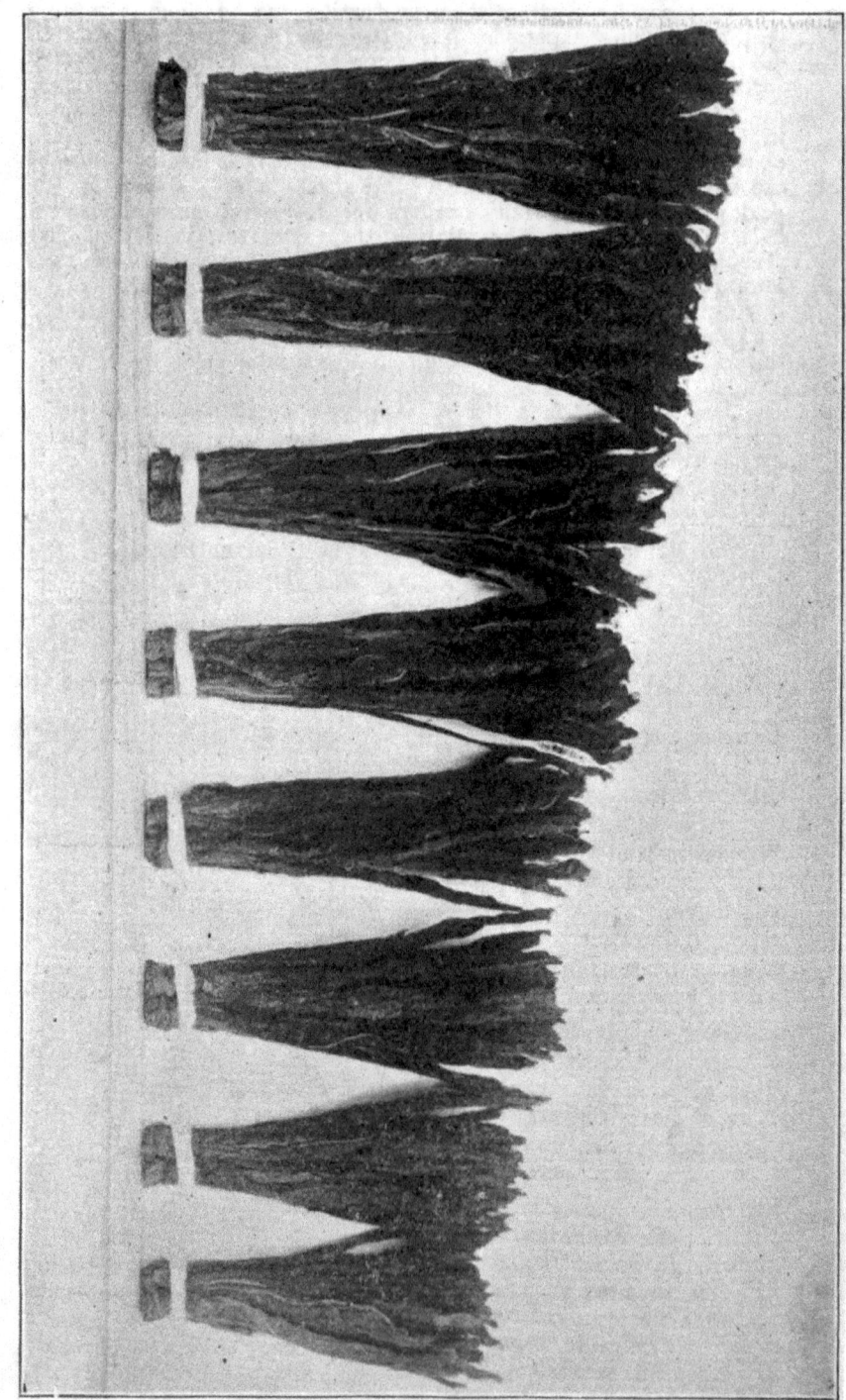

ZIMMER SPANISH; EIGHT GRADES, FROM ELEVEN TO EIGHTEEN INCHES (U.S. DEPT. OF AG.).

THE CULTURE OF TOBACCO.

Manufacturing and Export Types.

Bright lemon yellow, orange yellow and mahogany; grown in Virginia, North and South Carolina in various shades.
- Cigarette cutters.
- Cigarette wrappers.
- Plug fillers.
- Plug wrappers.
- Twist fillers.
- Twist.
- Lugs (wrapper).

Virginia, *Dark*.
- Snuffers.

Virginia, foreign types.
- English dark leaf for stemming.
- British olive green.
- Austrian, A, B, C.
- Italian, A, B, C, C.
- French, A, B, C.
- African.
- Spanish lug.
- Portugal cutting leaf.
- Export fillers.
- Export snuffers.
- Black fat.

Clarkesville types; grown in Clarkesville District, Tenn.
- Domestic, various sizes.
- Foreign, various sizes.
  - Italian.
  - German.
  - Belgian.
  - Austrian.
  - Spanish.
  - Swiss.
  - African.
  - British.

Kentucky, *Dark*.
- Domestic.
  - Brown snuffers.
  - Brown mottled.
  - Relandling.
  - Black wrappers.
- Foreign.
  - Italian A, B, C.
  - French A, B, C.
  - Spanish A, B, C.
  - African.

Burley; grown in Ohio and Kentucky.
- Red top fillers.
- Export lugs.
- Plug wrappers.
- Plug fillers.
- Twist wrappers.
- Twist fillers.

CONNECTICUT BROAD LEAF; DARK (U.S. DEPT. OF AG.).

Manufacturing and Export Types—continued.

Burley; grown in Ohio and Kentucky.
        Colory leaf.
        Cigarette leaf.
        Common lugs.

Smoking type; grown in Ohio and Maryland.
        Spangled.
        Fine yellow.
        Brown.
        Colory.
        Firm red.
        Seconds.

The Characteristics of a Good Tobacco.

At one time any tobacco that would burn was considered fit for human consumption, but along with the improvement and development of the plant, man's taste has been educated to the point of demanding certain essential characteristics in the tobacco intended for his use. There are many shades and variations in these characteristics depending on the trade catered for, because in their estimate of tobacco, as well as of wine, the best judges do not agree. However, certain points of excellence are always demanded.

If tobacco be intended for smoking it must have the ability to hold fire and to burn evenly, smoothly and thoroughly. It must not char, that is, there must be no black line between the ash and the unburned portion of the tobacco. If it be in the form of a cigar, the ash should be white and solid, and not flake and fall over the clothing.

Tobacco must have flavour. It must be sweet and pleasant, and not be too mild or too rank and strong. The flavour of the leaf must be agreeable and pleasant, for if the flavour is agreeable, the aroma of the burning tobacco is likely to be satisfactory. The aroma of a cigar is partly due to the volatilization of the products of the sweat, and partly to the destruction of certain compounds by a process of dry distillation. This process takes place largely in the interior of the cigar and in the heated portion near the coal. This distillation and volatilization create disagreeable odours as well as agreeable aroma, and a cigar can only be considered as good when the latter hides or subdues the former. This is true of cigarettes and pipe tobacco as well as of cigars, although in cigars the characteristics are more marked.

If a tobacco be intended for a plug wrapper it must have style, elasticity, toughness and body. Nor must it be too large or too small for the size of the plug to be manufactured.

If intended for a cigar wrapper the leaf must have style, and be elastic, thin in texture, finely grained, light and uniform in colour, and the stem and veins must be small and of the same colour as the leaf. The leaf should be as free from flavour as possible, as it is the portion that comes in contact with the mouth. The cigar wrapper can have much influence on the quality of the smoke for the reason that it is exposed to the air during combustion. The standard of excellence for wrappers is the Sumatra leaf, as the standard of quality in fillers is the Vuelta Abajo leaf.

32 THE CULTURE OF TOBACCO.

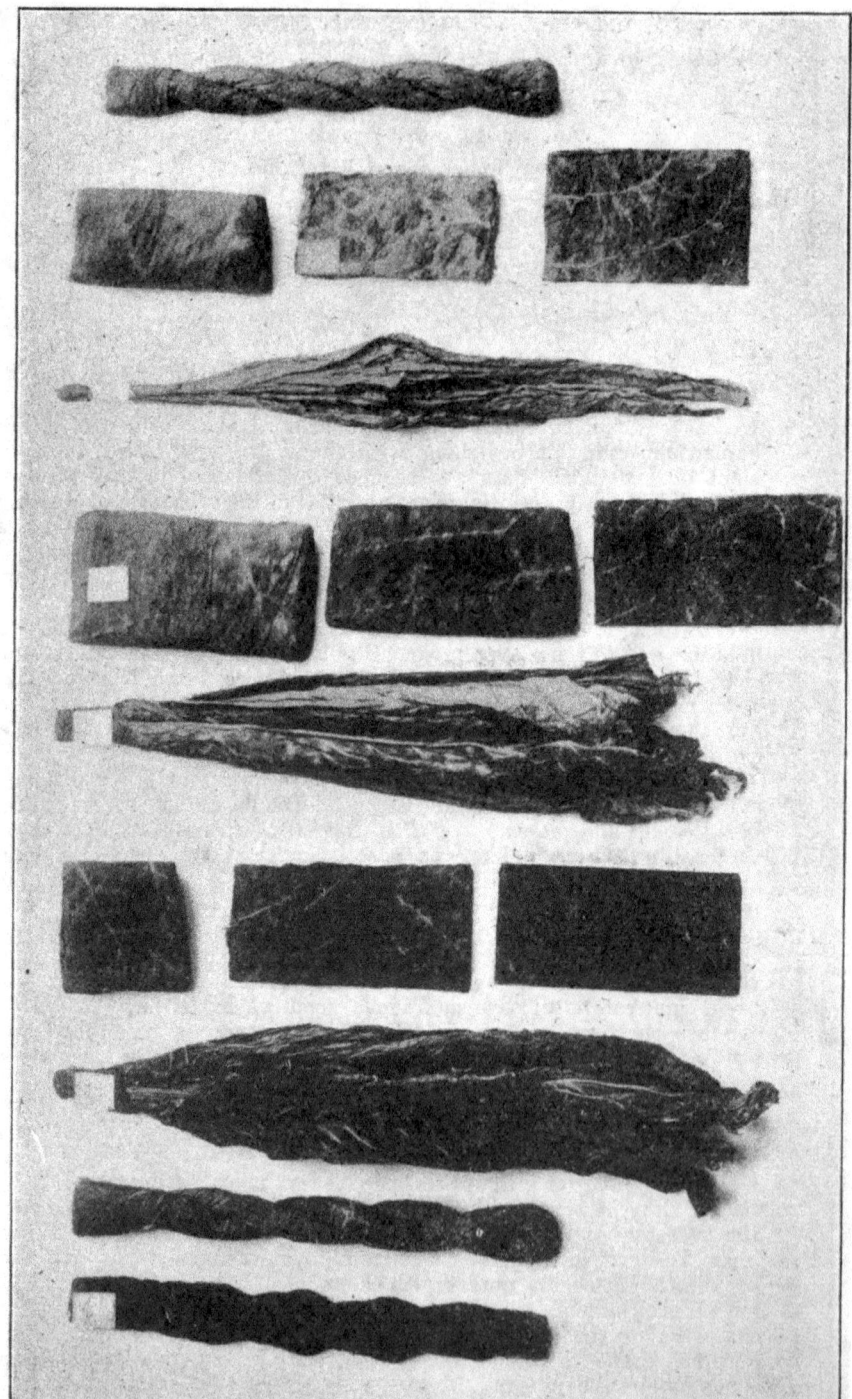

BRIGHT WRAPPERS AND STEPS IN THE OPERATION OF MAKING PLUG AND TWIST (U.S. DEPT. OF AG.).

If a tobacco be intended for chewing, it must have a certain toughness so that it will hold together while being masticated, and not break up into small flakes in the mouth. Chewing tobacco must also be rich in flavour. A tobacco that has a high absorptive capacity is eagerly sought after for this purpose, for the reason that large quantities of flavouring liquids and sauces are added. It is this ability to hold so large a quantity of flavouring sauces that gives to the White Burley its popularity.

Tobacco intended for pipe smoking and for cigarettes must be free from the gumminess so sought for in chewing tobaccos. This gumminess would interfere with the cutting or granulation of the leaf by machinery. The bright tobaccos are at the present time the fashion for cigarettes.

It is seldom that the desired standard of flavour is found in any one tobacco, and it becomes necessary to blend or mix different grades. Often this blending is done from the standpoint of economy, when a certain proportion of a perfect flavoured, but high-priced, leaf is used to give quality and character to a cheaper tobacco with all the requirements except the flavour. In some brands the Perique is largely used for this purpose. At other times the blending is conducted on the principle that, if but one tobacco be used, and at any time this particular tobacco becomes unobtainable, the substitution of an entirely different tobacco at such a time would ruin the established reputation of the brand by changing its character. Four or five, or even more, different grades of leaf are blended together, and the substitution of a different grade for any one of them does not radically change the character of the blend.

The colour of the tobacco demanded is largely subject to fashion, and may change at any time, but the aroma sought for remains the same. Smokers are more and more demanding a certain standard of excellence in this particular. Brain workers, and those living a sedentary life, prefer a mild tobacco, while those living a rugged out-of-door life—as sailors—seek a strong, stimulating tobacco. The colour of a tobacco is not necessarily indicative of its strength. A mild tobacco may be thoroughly fermented until it is dark in colour, while a strong tobacco may have the fermentation cut short and be light in colour. A heavy dark leaf is, however, likely to be rich in nicotine and other similar products. The colour of the wrapper on a cigar is not a good feature by which to judge the strength and quality of the smoke. Cigars are often encased in all colours of wrappers when made, and assorted according to colour later. A cigar should be selected by the sense of smell; a tobacco that has a pleasant aroma will usually be agreeable if the burning qualities are also good.

The palate and olfactories learn to accommodate themselves even to a poor smoke, and for this reason it is often a slow process to educate a nation to use a new good tobacco in preference to a badly-flavoured tobacco that has become established. A smoker may, because of his perverted taste, judge a bad tobacco to excel what is in truth a better tobacco.

The Tobaccos Called for by Different Countries.

The Germans demand a tobacco from 18 to 26 inches in length, with fine stems, sweet flavour, and strong and elastic fibre.

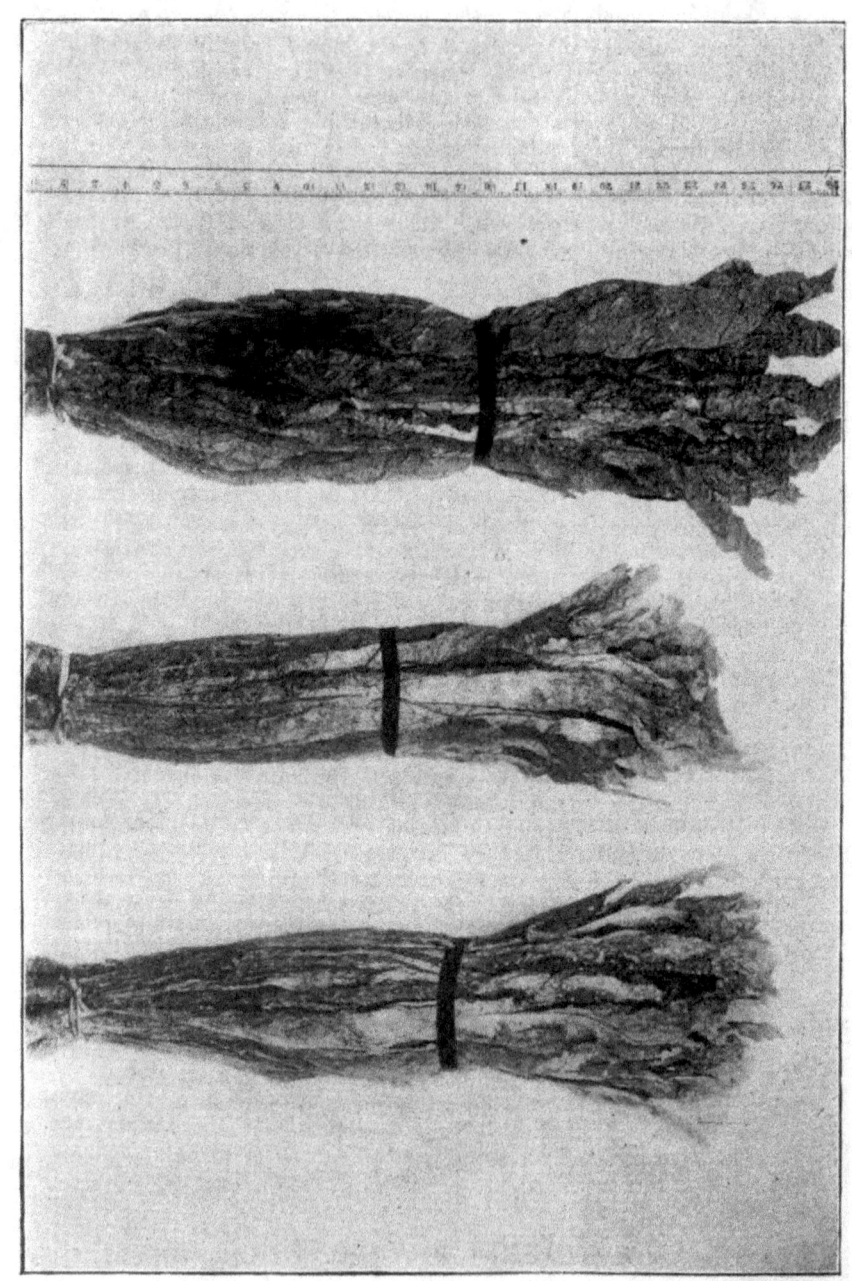

KENTUCKY DARK; GOOD LEAF, MEDIUM BROWN, GOOD LUG (U.S. DEPARTMENT OF AGRICULTURE).

The colour may vary from a cherry red to dark, the dark types being required for many purposes. Plenty of oil and of body in the leaf is insisted upon. Leaves that are mottled with yellow are used for treating with sauces and flavouring liquors before manufacture. This last type of leaf is called piebald, and in it the fibre must be yellow and the leaf black after treatment. The German "spinner," used for the manufacture of strand, must be of good length, tough, oily, fat, elastic, clear, and heavy. Germany purchases large quantities of tobacco for shipment to Russia and Scandinavia, and imports from the United States about 40 million pounds of leaf tobacco a year, as well as buying the bulk of the Brazilian crop.

Great Britain takes many grades very similar to those of Germany. Because of the high duty, all tobacco shipped to England is stemmed, and placed in as dry a form as possible before shipment. The very strongest of all the tobaccos, and the one with the highest nicotine content, is used for the manufacture of Navy plug. For the Bird's Eye cutter, the tobacco must be imported in the whole leaf, which must be of an even bright colour on both sides, smooth and clean, with but little oil or body. The upper and lower sides of the stem vary in colour, and thus give the bird's-eye appearance when cut. Great Britain is buying more and more of the better grades of tobacco, and is rapidly increasing the consumption of Bright leaf. The standard tobacco, however, is the heavily-smoked, slightly olive-coloured leaf. The taste for this creosotic-flavoured tobacco was formed in the earlier days, when it was necessary to have a smoked tobacco, in order to withstand the ocean voyage. This olive-coloured tobacco is harvested before fully ripe. Great Britain has a market for any tobacco that falls within her standard types, but is somewhat slow to take up with a new tobacco; however, there is an excellent market awaiting any new type of cigar leaf that may be produced. America sells to Great Britain nearly a hundred million pounds of leaf each year.

The Italian leaf should be smooth, silky, of good length, and of a slightly lighter colour than the German types. The leaf must also have less fat. The three types required vary in length from eighteen to twenty-six inches, and are used for cigars, cigarettes, and snuff. The tobacco is purchased and manufactured by the Government under what is known as the Regie system.

The Austrian types are firm, tough, elastic, glossy, of uniform red-brown colour, and of good length, and are chiefly manufactured into cigars.

The Swiss take but little tobacco, but buy the best leaf, and pay the highest prices. It is mostly required for cigar wrappers.

Spain will take any tobacco that is cheap enough.

France, as a rule, buys the poorer grades of tobacco, and is inclined to estimate the value of the leaf rather by the length than by other qualities, preferring an over-ripe leaf, which must be from reddish-brown to red in colour, and from eighteen to twenty-five inches in length. The same leaf may have several shades of colour, but must be clear and supple, without much elasticity or body.

The tobacco intended for North Africa is of medium length, and light or mottled in colour. For the Upper West Coast long, dark and strong; for the Lower West Coast long, medium or light in colour, with fine fibres. During the year ending June 30, 1902, the United States exported to Africa five and a half million

BURLEY FLYER; EXPORT LUG; DOMESTIC LUG (U.S. DEPT. OF AG.).

A LARGE PLANT BED.

pounds of leaf tobacco, three million pounds of plug, and three hundred million cigarettes; also a quantity of cigars. The most of the plug and cigarettes went to British South Africa, as did one hundred and fifty thousand pounds of the leaf. West Africa, and particularly British West Africa, consumed the major portion of the unstemmed leaf,

The United States annually exports about three hundred million pounds of leaf tobacco, at an average export price of fivepence a pound. This consists of nearly all of the lower grades of tobacco. Besides this, the United States manufactured about three hundred million pounds of pipe and chewing tobacco, fifteen million pounds of snuff, six billion cigars, and five billion cigarettes, as well as importing about twenty million pounds of foreign leaf, and quantities of cigars and manufactured tobaccos. As yet the United States has been an importer of the better grades of cigar tobaccos, but that portion of the industry is now being pushed in a scientific manner, and it may be only a short time before the United States will also be an exporter of Cuban fillers and Sumatra wrappers.

THE SEED BED.

The tobacco seed is very small, and the reserve material for the nourishment of the young plant is soon exhausted. As a result the young plant is forced to prepare its own food much sooner than is the case with most plants. Because of this the young plant makes a very slow growth in its initial stages, so that soil and plant food must be placed in as favourable a condition as possible to aid the young plant through this critical period.

If all the seeds were fertile and capable of germination, one ounce of tobacco seed would be sufficient for three hundred thousand plants. Experience has shown, however, that at least seventy-five per cent. of the seeds are sterile, and that many of the remainder will produce small and unthrifty plants, so that for every thirty thousand plants required it is necessary to allow one ounce of seed. Thirty thousand plants would be sufficient to set two acres of the cigar tobaccos and from four to seven acres of the other types of tobacco.

The plant bed for one ounce of seed should cover about fifty square yards. Tobacco seed is not expensive when the area that a small amount will plant is considered, and it is far better to have too many plants than not to have enough or be forced to set weak plants. Many planters sow several times as many seed beds as would be necessary under ideal conditions, and allow short intervals of time to elapse between the sowings of the different beds. This assures them a sufficient quantity of healthy plants even if there be some losses. Then the time of planting varies with the different seasons, and it is an important matter to have plants of just the right size to set out when they are required.

Before sowing the seed it is a good plan to test the germination. A hundred seeds are counted out and placed between two pieces of moist blotting paper. The paper is kept between two plates and at a temperature of from seventy to eighty degrees for ten days, at the end of which time the percentage of good seed is determined by counting the number of seeds germinated. This will enable the planter to make up for any deficiency in germination by the use of a

ONE TENTH OF THE PLANT BEDS OF A TEXAS TOBACCO PLANTATION.

larger quantity of seed. It must be understood that the conditions in the soil are not so favourable to a high percentage of germination as is the blotting paper.

In choosing the location of the plant bed, there are several points to be considered. First, a good, rich, friable well-drained soil must be chosen. Next, a location that is protected from the prevailing wind and on an exposure where the bed will receive all the sunlight possible. In America these beds are usually made on the southern slope of an open space in the forest.

American growers give the site of the seed bed a thorough burning. The object of this is to destroy all insects, or their larvæ, that may be in the soil. In order to accomplish this burning, the soil is first laid with poles in order to keep the burning wood off the soil and admit the air. Upon these poles the wood is piled and the fire started on the leeward side that the progress may not be too rapid. A slow fire will convert all the moisture in the surface soil to steam, and thus cook anything that may be in the soil. The fire is continued long enough to convert all the moisture in the first three or four inches of soil to steam. The use of steam instead of fire has been tried for the preparation of the plant bed and has proved very satisfactory. The steam does not destroy the combustible portions of the soil while it cooks and destroys all seeds and insects. The steam is applied by running it through a pipe or hose and confining it under an inverted wooden or metal case. A packing case that has been made steam tight will answer for this purpose. The use of steam is to be recommended where the planter has a traction engine at his disposal.

After the bed has been thoroughly burned the soil is broken up to a depth of two or three inches and all roots and other trash carefully removed. Care must be taken that the soil be not dug too deep, for this would bring to the surface weed seeds that have been buried too deep to be destroyed. Fertilizer is thoroughly worked into the soil. About three pounds of any good commercial fertilizer to each ten yards of the bed will answer the purpose. Nitrate of soda is the best fertilizer with which to give the plant a quick start and rapid growth, but this chemical should be used more sparingly than the other fertilizers. Barn yard manure would be excellent were it not that it usually contains quantities of weed and grass seed that would neutralize all the good results of the burning.

The seed should be sown about sixty days before planting time. To sow the seed mix it with ashes or fine meal, at the rate of one-fourth ounce of the seed to two quarts of the ashes or meal. The object of the ashes is to give greater bulk, thus enabling the planter to sow more evenly. The colour of the ashes or meal also informs the sower whether all portions of the ground have been sown with a sufficient quantity. After being sown the seed is not raked in, but the surface of the soil is lightly brushed with a broom. The sprinkling of the bed with water from an ordinary gardener's can will in itself be sufficient covering. Tobacco seed that has been buried too deep will not germinate.

The plant bed should be covered with a light cloth or muslin. This is stretched over a framework of boards or bricks that has been built around the border of the bed. Small sticks about six inches high are set up throughout the bed, to keep the canvas from touching the plants. This canvas will shelter the plants from the intense heat of the sun, and will at the same time retain during the

SHOWING THE ROOT SYSTEM OF A TOBACCO PLANT.

night a portion of the heat of the day. It will also protect the plants from drying winds, frosts, and insects. Where the cloth cannot be secured a light covering of grass, or even of light brush, will serve to partially protect the bed from the sun, winds, and frost. This covering should be raised at least a foot off the bed, by means of a light framework.

Unusual care must be exercised in handling the plant bed, for it is a very easy matter to so injure the plants as to dwarf them throughout their life period. The soil should be kept moist, but not wet. If too much water be used the plants are not only likely to be smothered, but conditions are made favourable to the development of fungus diseases. Within two weeks after sowing the plants will appear; for some time it may appear that they are not making any growth; this is the period when they, having exhausted the material stored in the seed, are adapting themselves to the new conditions of life. In a month after sowing a rapid growth will start, and in another month the plants should be ready for transplanting. Plants are better if not allowed to become too large before planting, as an over large plant is more likely to grow lanky. American growers prefer to set out the plants when the top covers about the area of a five shilling piece. During the last month in the plant bed it will pay to water occasionally with liquid manure. For a week or more before planting, the canvas covering should be removed, to allow the plants to become hardened. If the plants appear to be growing too thickly they should be thinned out; one good plant to each square inch is sufficient. This will give about thirteen hundred plants to the square yard.

In drawing the plants for transplanting care must be taken to secure as much root as possible. The plant bed should be thoroughly soaked, so as to allow the roots to be drawn with little breakage. When the large plants are drawn out of the bed the soil should be immediately watered, if not already done, in order to pack the earth around the roots of the remaining plants, which will have been disturbed by the operation. As soon as the plants have been drawn they should be placed in a basket, with the roots downward. The top of the basket must be covered with a cloth, and the basket put in a cool place until the plants are set out.

PREPARATION OF THE LAND, AND PLANTING.

Within sixty days after it is set in the field the tobacco plant has reached maturity. In order to properly accomplish this result, the soil must be well tilled, so as to allow the roots to rapidly pursue their search for nourishment. The soil must be in a condition to admit air to the rootlets and also to the beneficial nitrifying bacteria in the soil, and at the same time it must have the ability to hold sufficient moisture for the needs of the crop. This condition can be reached only by a thorough preparation of the soil before planting time. The fibrous rooted tobacco plant seeks its food near the surface, but because of this fact the subsoil should not be neglected, for on the condition of the subsoil depends largely the moisture content of the surface soil. The land should have one, and, if possible, two thorough ploughings, and then be placed in as fine a tilth as possible by the use of harrows.

There are two methods of growing the plants: one method is known as level cultivation and the other as hill cultivation. The system of level cultivation is perhaps best where there is no danger of excessive moisture in the soil. In this case the field is

marked off into rows, three to three and a half feet apart, and the plants set at the distance of three feet in the rows. If it is desired to use the horse cultivator in both directions, the field is marked both ways, so that the plants will "check row." If the soil is inclined to be wet at any time during the season, the field is thrown up in ridges three and a half feet apart, and the tobacco set on the tops of these ridges. The ridges are formed with a small plough or with a horse cultivator using the wing adjustment.

The distances of three and three and a half feet hold true only for the ordinary smoking and manufacturing types of tobacco. Most of the cigar varieties are planted much closer together in the row. The Cuban and Sumatra tobaccos are but a foot to fourteen inches apart in the row. Some of the stronger growing varieties are planted from eighteen inches to two and a half feet apart. The distance depends on the size of the plants. If a cigar tobacco be given too much room, the leaf will become too large and coarse for cigar purposes. Sumatra tobacco when first grown in Florida was very nearly a failure, for the reason that it was set too far apart and became coarse. In Sumatra, where hand labour alone is used, the plants are placed two feet apart in each direction; but Sumatra has a heavy rainfall and a tropical climate, so that the plants grow very rapidly and are fine in texture. This distance might not do for localities where the condition for a rapid growth was less favourable. In America the rows are always at least from three feet to three and a half feet apart, for the reason that most of the work is done by means of horse cultivators. In a new locality growers must determine the distance by experiments.

If the day be cloudy planting may be done at any time, but if the day be hot and dry, the planting should be left until the last half of the afternoon, so as to give the plants an opportunity to recover and establish themselves during the coolness of the night. Where the planting is done by hand, the plants are dropped along the row at regular intervals by a small boy. Immediately after the plant

44 THE CULTURE OF TOBACCO.

A "BEMIS" TRANSPLANTING MACHINE.

distributer there follows a man or boy, who, with a round stick about ten inches in length, makes a hole, and then inserts the plant into the hole, and, while holding it firmly with one hand, presses the earth firmly around the roots with the stick. The surface of the soil is then rapidly smoothed over and left in as loose a condition as possible. When the soil is dry or the weather unfavourable, one person carries a pail of water, and either goes just before the planter and pours a little water into the hole or follows close after, and places the water in a small hole made beside the plant. If this additional hole is made, it must be covered up as soon as the water soaks in to prevent evaporation. Even if the soil appear to be sufficiently moist the use of water in planting will give good results, for the reason that it settles the earth firmly around the roots of the young plant, and thus permits it to start growth at once. A man and a boy will plant about five thousand plants a day, and one extra person can do the watering for that number of plants, while a fourth person with a wagon can haul the water for a large number of planters.

In some sections transplanting machines alone are used. These machines set the plants at the desired distances and waters them at the one operation. The machine will also, if desired, place a small quantity of fertilizer with each plant. Three persons are required to handle such a machine, one to drive the two horses, and the others to feed the machine with the plants. Five acres of ordinary tobacco may be thus planted in a day.

In Cuba the plants are not set in holes, but a small furrow is made with a shovel cultivator or a small plough. The plants are then rapidly set by being placed in this furrow and a handful of earth drawn to each. The soil is then levelled up with a hoe. This proves to be a speedy method. The furrows should not be made much ahead of the planter and must be filled up at once so as to avoid evaporation.

CULTIVATION.

The soil is supposedly in fine condition and tilth when the plants are set, and the aim of cultivation should be to keep it in that condition. The tobacco plant is largely a surface feeder, and its roots do not penetrate deeply in the soil. For this reason, shallow cultivation only should be used, so as not to cut off the roots and check the growth of the plant. Cultivation is not merely for the purpose of killing the grass and weeds, but also for the admission of the air necessary to the roots of the plants, as well as to the nitrifying bacteria. After wet weather, cultivation hastens the drying out of the soil, and, in dry weather, cultivation by the creation of an earth mulch prevents excessive evaporation of moisture. No set number of times can be given for the cultivation of the field, but it should be done whenever the soil begins to harden or crust over. The best implement for the purpose is an adjustable horse cultivator, or horse hoe, as it is sometimes called. This implement is easy to handle and can be adjusted so as to throw the earth to, or away from the plant as it may be desired. When the plants are small, a wheeled riding cultivator may be used that will cultivate both sides of the row at the same time; two horses are required for this implement. As the plant grows larger the earth near the plant should be stirred with a hoe, and later when the leaves of the plant have become so large that there is

46 THE CULTURE OF TOBACCO.

danger of injury to the plant by the use of the horse, the hoe and hand labour can alone be used. No other plant will respond so readily to cultivation, nor will the value of any plant be so easily affected by neg-

lect as the tobacco plant. A rapid growth from the time of planting until the ripening period will give that fine even texture so desirable in high class tobacco. When the plant has commenced to ripen, cultivation should cease, for any further stirring of the soil will tend to induce an injurious second growth.

TOPPING.

The tobacco plant has a tendency to the production of seed, and for the reason that the seed is formed at the expense of the leaf, the blossom or terminal bud must be removed, and with the blossom must be removed all the leaves in excess of those that the plant can properly develop and ripen. The time at which this operation of topping should be performed depends on the use to which the cured tobacco is to be put, as well as to conditions of soil and climate and the individual characteristics of the plant. Strong, vigorous plants are topped high and given more leaves to develop than the less vigorous plants. Expert tobacco men do not count the number of leaves to be left on a plant, but through long practice are able to top the plant to the right number of leaves for its capacity. The number of leaves that a crop will average will depend largely on the season; one year but ten leaves may be left to a plant, and the following season an average of sixteen leaves may be left and the product be of equal value both seasons. The removal of the terminal bud produces a great change in the plant by increasing the surface, and thickening the leaf. The operation also causes an increase of the protein compounds and nicotine in the leaf as well as hastening the process of ripening. If the plant is topped too low there will be a rapid thickening up and curling of the leaf; if this occurs some of the top suckers should be allowed to grow and check this tendency. If a plant is topped too

soon there will be a loss of aroma in the cured product. The majority of the flower buds should be allowed to appear before the crop is topped, so as to permit topping to be completed in one operation.

Priming.

At the time of topping many growers pull off the lower leaves of the plant, and this operation is known as priming. These leaves are usually small, thin and very dirty, and would not produce saleable tobacco even if left to mature. The exponents of priming maintain that these leaves being useless should not be allowed to appropriate the sap that would otherwise go to the other leaves. It is contended that these "sand lugs" furnish a harbour for insects, and also that the removal of these leaves will leave more room for the use of the hoe. Other growers leave these "sand lugs" upon the plant with the thought that they protect the next higher leaves from being spattered with soil during rains.

Suckering.

As soon as the plant is topped, or even before, small shoots or suckers start out from the axils of all the leaves. If these suckers be allowed to grow they will greatly lessen the amount of tobacco produced and also injure the quality. As soon as the suckers appear they must be broken off. This will have to be repeated at least once a week. The work can easily be done by children, but care must be exercised that the remaining leaves are not broken off or injured. In some sections of the world, one of the lower suckers is allowed to remain and produce a second crop after the harvesting of the first. This sucker crop is inferior in quality and does not pay, for the reason that it lowers the standard and reputation of the tobacco of the district.

During the whole period of growth the tobacco plant is subject to injury by insect pests, and against these perpetual warfare must be waged. (See the chapter on insect pests and their treatment, page 63.)

RIPENING.

Ripening is hastened by the topping and suckering operations. The remaining leaves are filled with an abnormal accumulation of organic compounds that would have been used for the development of seed and other leaves had the tobacco plant been allowed to mature normally. The plant at this stage largely increases its percentage of acids, nicotine and protein compounds. A ripe leaf has a rough

feeling to the touch, and when folded between the fingers will easily crack. There will also be a change in colour from a dark to lighter shades of green, and the appearance of yellow spots. This indicates the maturity of the leaves and the translocation of material from the older portions to the less mature leaves of the plant. A brownish colour may also appear around the borders of the leaf. A leaf if

STRINGING LEAVES ON STICKS TO BE FLUE CURED.

harvested too green will always have a tendency to retain a greenish shade, and will be deficient in grain and slightly bitter when smoked. There is also such a thing as an over ripe leaf. Over ripe leaves contain more water and less organic material than they did when at the proper stage for harvesting. This is due to the fact that in the ripening leaf the chlorophyll grains gradually change to other forms, and thus cease their function of forming new organic matter, while at the same time consumption of the already stored material continues. An over ripe leaf will seldom cure up an even colour and will be brittle rather than elastic.

The ripeness of shaded tobacco cannot be determined by the same indications as can the sun-grown crop. A general appearance of maturity is the only guide that the planter will have.

HARVESTING.

When the proper stage of ripeness has been reached the time has arrived for the harvest. All portions of the tobacco plant do not ripen at the same time, and because of this fact two different methods of harvesting have been developed. In the one system the whole plant is harvested on the stalk when the middle leaves of the plant are mature, while in the other system each leaf is primed off as it becomes ripe. The first system is accomplished with the minimum of labour, although it will not produce so large a percentage of properly ripened tobacco as the second system, for when the middle leaves are at the proper stage of ripeness, the lower leaves are over ripe, and the top leaves still green. The advocates of the stalk cutting or whole plant system maintain that a large percentage of the material in the stalk is transferred from the stalk to the leaf during the process of curing, and this claim seems to be substantiated by experience. This system is the one largely used for the bulk of the tobacco in America, but the finer grades of cigar leaf and a portion of the Bright tobacco are harvested by the single leaf method. In some places a combination of the two methods is adopted, the lower leaves being harvested singly as they ripen and the upper half of the plant taken off with the stalk. Where the leaves are primed they are at once placed in baskets and hauled to the curing barn. At the barn these leaves are strung on twine by means of a needle run through the end of the stem, or else the twine is looped around the end of three or four leaves at a time and the bunches of leaves left several inches apart. This twine is then fastened to a four foot stick, and the stick is hung on the tier poles in the curing barn. The aim is to get about forty leaves to each stick.

When the stalk system of curing is used, the plants may be cut off at the surface of the ground with one stroke of a knife, and then strung on the four foot curing stick by means of a detachable iron spear head that is fitted into the end of the stick. Another method is to, with one stroke of the knife, split the plant from the top nearly to the ground, and then to sever it below the end of the vertical cut with a second stroke. The plant can then be placed astride the curing stick. The split plants will cure out more rapidly than the speared plants, but will suffer a slightly greater loss of weight. With small plants, as the Cuban tobacco, the stalk is too small to stand much splitting, and is speared into the stick.

HANGING LEAF IN A CONNECTICUT TOBACCO BARN (U.S. DEPT. OF AG.).

LOADING TOBACCO ON WAGON IN SOUTH CAROLINA.

TOBACCO CUT AND TURNED UP TO WILT.

54 THE CULTURE OF TOBACCO.

TOBACCO CUT AND HANGING ON THE CURING STICKS TO WILT.

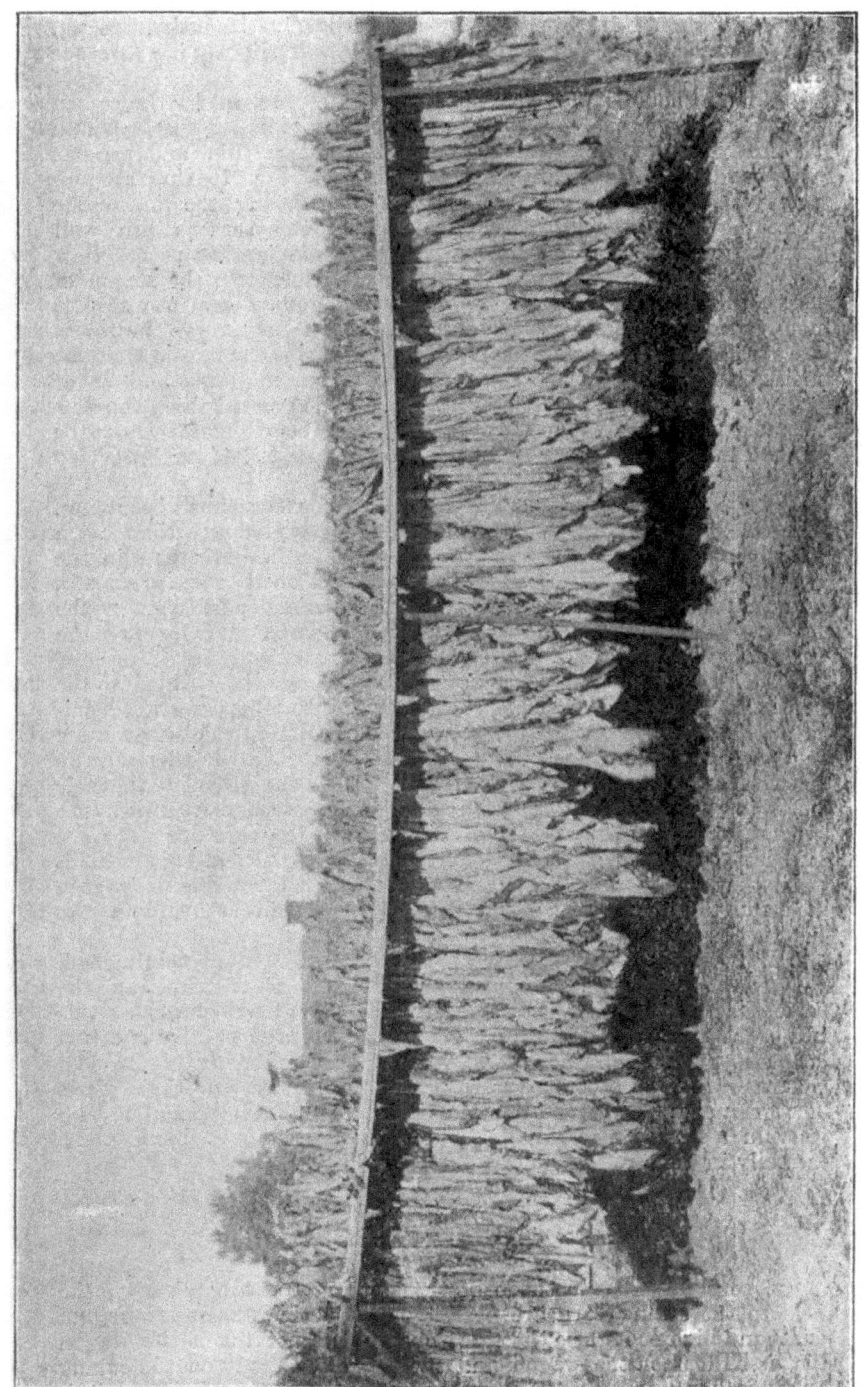

TOBACCO WILTING ON RACK IN FIELD, TENNESSEE.

Cigar leaf should continue as long as possible in the curing process. The unsplit stalks being slower in drying will slightly increase the length of the curing period.

Plants should not be harvested soon after a rain, for the reason that the water has washed out the gums and oils of the leaf, and the leaf, if harvested at that time, will cure up thin and papery, and be devoid of the finer aroma. In fact, the ripening tobacco plant would be better off if it had no rain during the few weeks preceding harvesting. In countries where the seasons are well defined, growers should so arrange the time of planting as to allow the crop to ripen, and be harvested, before or after the season of heavy rainfall. In Cuba this point is understood and the crop is not grown during the heavy rainy season. Heavy dews, however, are greatly desired during the ripening period, for the reason that the moisture on the leaf aids in the formation of gums and other aromatic materials. Plants should never be harvested when the dew is upon them, for at this time they are very brittle and likely to be broken, and at the same time the dew covered leaf will develop black spots when cured.

Bright sunny days should be chosen for the time of cutting. The acidity of the leaves is less on a warm sunny day than upon a cool or cloudy day, and is also less in the evening than in the morning. On a bright day the processes of metabolism are promoted, and the respiration stimulated, so that at the end of such a day the leaf will have more of the desirable products and less of the undesirable by-products than at other times.

The plant should be slightly wilted before being hauled to the curing barn, as this will prevent breakage and hasten the commencement of the curing process. Often the curing sticks are stuck into the soil at an angle approximating 65°, with the slope away from the sun, and on these sticks the plants are placed with their butts toward the sun and allowed to wilt for a short time. At other times the plants are placed with the top side down, or are laid in rows of three or four plants deep and allowed to remain until they are wilted. Long wilting is not so popular as it once was, and the system of placing the cut plants on scaffolds in the field is obsolete.

Leaves if left long exposed to the sun will become sunburned, that is the heat of the sun will kill the plant cells and at the same time destroy the enzymns that bring about fermentation; this being the case no further life changes or processes of fermentation can take place, and, with the exception of the drying of the leaf by the evaporation of moisture, no improvement will take place. The greenness of a sunburned leaf will always remain and reduce its value. Sunburned leaves will carbonize and not burn satisfactorily in the pipe or cigar.

TRANSPORTATION.

Several different methods of transferring the tobacco from the field to the curing barn are practised. Sometimes a frame four feet wide, and high enough so that the tobacco will not touch the wagon box, is built upon the wagon, and the tobacco, already on the curing sticks, is hung on this. This method is to be preferred where it is

WAGON, WITH RACK FOR TOBACCO.

58   THE CULTURE OF TOBACCO.

LOADING TOBACCO ON FARMER'S HANDY WAGON (PHOTO LOANED BY F.H.W. CO., SAGINAU, MICH.).

greatly desired to have no broken leaves, as is the case with tobacco used for plug or cigar wrappers. The reason for having the frame but four feet wide is that this is the length of the curing sticks. Sometimes the frames are built of double height so that two tiers may be hung. In Africa, where heavy ox teams and large wagons are in use, there would appear to be no reason why these frames should not be constructed of double width as well as double height.

Sometimes a low frame is made of not more than a foot in height, and on this the sticks are hung, and the plant laid out flat in the wagon. The plants are laid one upon the other after the manner of shingling or tiling, and the only advantage of the frame is that it prevents the butts of the stalks from injuring the leaves underneath.

At other times the sticks of tobacco are merely piled on the wagon with the butts of the stalks to the outside, and the tips interlacing. A very low form of a wagon with wide tires and a large top that extends out over the wheels is now becoming popular. It is of light draft and also does away with high lifts. If but little tobacco is grown, and that little in the neighbourhood of the curing barn, sleds or stone "boats" may be used for transporting the tobacco.

Where the leaves are primed they are placed in oblong baskets, and these baskets hauled to the curing barn as soon as possible. The stringing of the leaves is done under shelters built around the curing barn.

THE GROWTH AND SELECTION OF TOBACCO SEED.

When it is found by a grower that his district is particularly adapted to the production of a certain type of tobacco, he should aim to perpetuate or even improve that type by the growth and selection of his own seed.

The best plants in the field should be selected at the topping time, and the flower buds of these plants left to mature seed. It is advisable to first make a preliminary examination of the field, marking those plants that more nearly fulfil the conditions sought, and then to go over the marked plants several times, and by a process of elimination of the worst specimens, reduce the number left for seed to the few best plants. In selecting the plants, a certain ideal must be held in mind, and the grower rigidly judge all the candidates by the standards of that ideal. If the grower depart from the standard in one particular, for some plant that seems unusually fine in some one point, and then modify his standard in some other particular for another plant, the result will be that the crop grown from the seed will be irregular in quality, and without uniformity. The ideal must be developed in the mind, and the standard established, before the grower goes into the field to accomplish the work of selection, and the result of the work when completed must be, that the plants left shall have a common resemblance. Selection rigidly carried out for one year will accomplish more in the improvement of the type than a slightly less rigid selection will accomplish in several years. The selection process carried out year after year will work wonders in the improvement of the crop. This point must, however, be regarded

A RHODESIAN FIELD OF TENNESSEE EXPORT TOBACCO LEFT TO MATURE SEED.

carefully, viz., that the ideals striven for must not be radically changed with each year. If later it be found that the ideals of the first years have been incorrect, then, of course, it is necessary to change them, but changes should not be made according to the whims and fashions of each year. Characteristics established by a selection of three years are at least ten times more permanently fixed than those established by one year's selection. There is always a tendency in all highly-developed plants to revert to type, and this tendency is stimulated by a soil or climate that is in any way different from the plant's native home. The object of selection is, therefore, not only the improvement of the type, but its preservation.

Seed should not be saved from different varieties of tobacco grown in the same locality, for the pollen is likely to be carried from one variety to the other, thus producing a cross, so that the plants grown from the seed will largely differ in their characteristics. Some will resemble one parent and some another; some will have the characteristics of both parents in different degrees, and others will have none of the varietal characteristics of either. Where seed is saved from different varieties grown on the same farm, these varieties should be widely separated from each other. When several planters in the same locality each grow several different varieties of tobacco, they may often arrange each to save seed of but one variety, and to exchange each year.

The tendency to cross fertilization, and its resulting variations, is valuable where the process is directed by the hand of man, for it is often necessary to produce a new variety with the characteristics of two or more other varieties. When this is attempted, the two plants that are to be used are carefully selected, with the end sought in view, and the blossom bud covered with a small paper or muslin sack. Before the blossom has matured, this sack is lifted, and the pollen-bearing portion of one blossom (the anthers or male organs) is removed with a small tweezer; the sack is again placed on, and when the blossom is sufficiently developed, and the pistil receptive, the pollen is taken from the other plant, and placed on the blossom of this plant by means of a camel-hair brush, after which the sack is again adjusted, and left on the plant until the seed pod has formed. The seed from this plant is sown by itself, and the plants resulting show hundreds of varied characteristics and differences. These plants are gradually selected, and cut out of the competition, until only the two or three plants conforming nearest to the ideal are left to develop seed. The process of selection must be kept up for several years before a new variety may be considered as established. The results of the cross may not prove at all satisfactory, and the process may have to be repeated several times.

The establishment of a seed plot is an excellent scheme. The seed for the planting of this crop should be saved from the most perfect plant of the preceding year. Thus the best plant of the seed plot will be the parent of the seed-bearing plants of the following year and the remainder of the selected plants of the seed crop will be the parents of the main crop the following year. In this way a pedigreed variety of plants will be established and the plants of the seed crop will always be the offspring of the same parent and the plants of the main crop will always be the descendants of the same grandparent.

Theoretically, this constant inbreeding would in time lessen the vitality of the variety, but in practice this result will rarely be reached, for the isolation of the seed plot will seldom be so complete that there will not be an occasional accidental introduction of new blood.

When a plant is left for seed the top leaves are stripped off so as to reduce its top heaviness. If the plant shows too great an inclination to yield to the wind and be whipped around, it is steadied by being tied to a pole set in the ground. No suckers or secondary blossoms are permitted to grow, and as soon as the best seed pods have commenced to ripen, the remaining pods and blossoms are removed, the idea being to have a few well-nourished seeds rather than a large number of poorly nourished ones. When the pod is ripe the seed head is cut off and hung in a dry room, all imperfect pods being rejected in the operation. Later the seed is broken out of the pod and carefully freed from all dirt and chaff by winnowing. It is then placed in glass jars or bottles and sealed. If carefully kept in jars, tobacco seed will retain its vitality for at least ten years. Before being placed in the jars the seed must be absolutely dry.

Many tobaccos, like the Cuban, will not produce a large crop if grown from the carelessly produced imported seed, and, for this reason, it is the custom in many places to grow the seed of such varieties for years before saving seed for the main crop. However, if this process be kept up for any length of time the product will have a tendency to greatly deteriorate from its original qualities even though it has become acclimatised and increased in hardiness and productiveness. For this reason about the third or fourth year after the importation a large quantity of seed is saved and stored in airtight jars to be used for the production of the crop during the next eight or ten years.

Tobacco seed may be worth anything from four shillings a pound for the commoner varieties, to four shillings or even twice as much per ounce for selected cigar tobaccos. The average price is about twelve shillings a pound for the ordinary varieties and thirty shillings a pound for cigar varieties. An ounce may contain anything from two to four hundred thousand seeds, so that high priced seed is not necessarily expensive when the area that it will plant is considered.

Tobacco seed as a rule should not be purchased from general seedsmen, but should be secured from men who make a speciality of growing tobacco seed alone. From one to two hundred pounds may be produced to the acre where the selection of seed is not carried out very thoroughly, but it is much better to purchase seed at ten times the ordinary price from growers who select until they only produce one-tenth as much seed. Quality rather than quantity should be the object.

INSECT PESTS OF TOBACCO.

From the seed bed to the curing barn the grower must wage one continual war against insect pests. Many of these would entirely ruin the crop if left to themselves. To-day, through the use of arsenical poisons, it has become a simple matter to hold in check the once much dreaded enemies of the tobacco field.

Young plants grown under canvas and on burned soil are seldom subject to the attacks of insects.

SPRAYING WITH A KNAPSACK SPRAY PUMP (U.S. DEPT. OF AG.).

"**Cut Worms.**"—Usually the first pest from which injury may be expected is the "cut worms." These are the larvæ of several different species of moths and do great injury to all forms of tender plant

life by severing stalks near the surface of the soil. Several species also climb the plant and feed on the leaves. When a field is known to be affected with these pests it is the custom in many localities to scatter over the field little bunches of green grass, or other forms of vegetation, that have previously been sprayed with Paris green and water. In the absence of other food the pests will feed on this poisoned vegetation and be destroyed before the tobacco has been set on the field.

Another method is to scatter poisoned meal throughout the field. This bait is prepared by mixing one pound of Paris green, or of arsenic, with from fifty to one hundred pounds of mealie meal (ground maize or Indian corn meal), the meal being moistened with a little water and molasses. This poisoned meal is then placed in rows here and there throughout the field and among the growing plants. The cut-worm prefers the poisoned meal to the green vegetation and thus comes to an untimely end. The most satisfactory results are, however, obtained if the meal be placed in the field before the plants are set.

"**Horn Worms**" or **Caterpillars.**—These are the larvæ of Sphinx moths, and the name of "horn worm" is due to the fact that they have a small stout horn attached to one of the posterior segments of the body. These caterpillars have insatiable appetites, and two or three will ruin a tobacco plant in the course of a day; nor are they fond of the tobacco plant alone, but feed as well on other forms of the Solanaceæ as the tomato and potato.

Two different species destructive to tobacco are common in America, Protoparce Carolina and Protoparce Celeus. The Rhodesian tobacco caterpillar may prove to be a distinct species, but that fact will make no difference in the treatment to be followed as the habits are the same.

The mother moths may be seen and heard flying around the Petunia and Datura blossoms just at dusk. It is at this hour that the eggs are laid. In four to eight days the young caterpillars hatch and start their destructive career, which may continue for a month, at the

66 THE CULTURE OF TOBACCO.

SPRAYING DRY POISON ON THE PLANTS TO DESTROY INSECTS.

end of which time they have reached their full growth, and are prepared to burrow into the soil and transform to pupæ.

Where this pest is not very abundant, the method of sending children over the field to destroy all the caterpillars that can be found is a very satisfactory one. But where the pest is abundant, the field should be sprayed with a preparation of Paris green. One pound of Paris green to one hundred and fifty to two hundred gallons of water is about the right proportion. This preparation should be evenly distributed over the plant by means of a spray pump with the nozzle adjusted to throw a very fine misty spray. The Knapsack spray pumps are well fitted for this work.

In some localities a hand machine is used, which, by means of a rapidly moving fan, blows the pure Paris green evenly over the plant.

With this machine, as well as with the spraying preparation, care must be exercised that not too much of the poison be used, or the plant will be injured. The poison in no way injures the tobacco for human consumption, as it has been demonstrated by experiments that the poison practically all disappears before the tobacco is cured.

In many localities a method of destroying the moths themselves before they lay their eggs is practised. This is easily done, because of the habit that the moth has of feeding on the nectar of the Datura Stramo (called Jimson or Jamestown weed in America). Quantities of this weed are allowed to grow around the edge of the tobacco field, or the plants may be even set here and there throughout the crop. Every evening the blossoms are poisoned by having a few drops of the following mixture placed in them:—cobalt, one ounce; molasses or honey, one-fourth pint; water, one pint. The moths die very shortly after partaking of this preparation. When this method is used the Datura should not be permitted to ripen seed, or else it will develop into a worse pest than the caterpillars.

68   *THE CULTURE OF TOBACCO.*

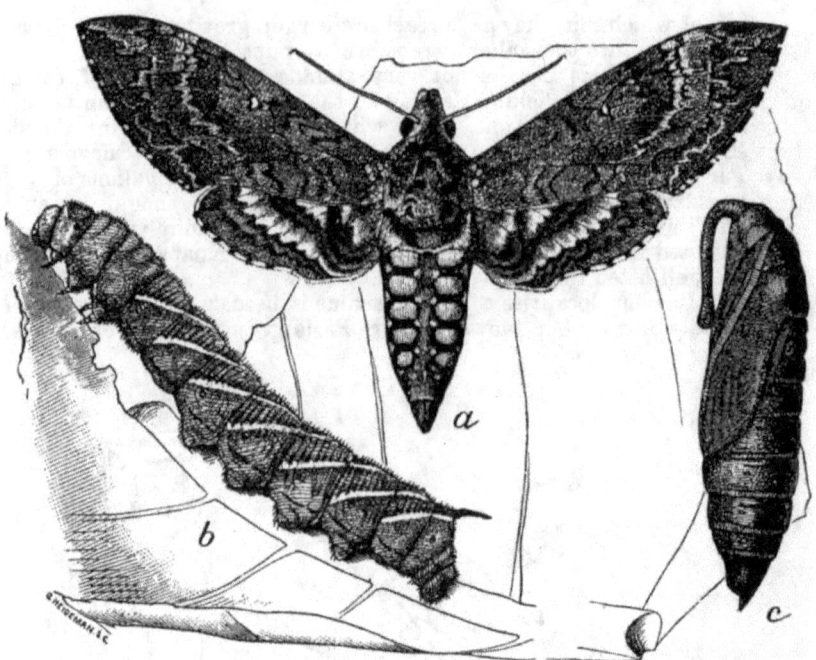

FIG. 5.—Southern tobacco worm (*Protoparce carolina*): *a*, adult moth; *b*, full-grown larva; *c*, pupa—natural size (original).

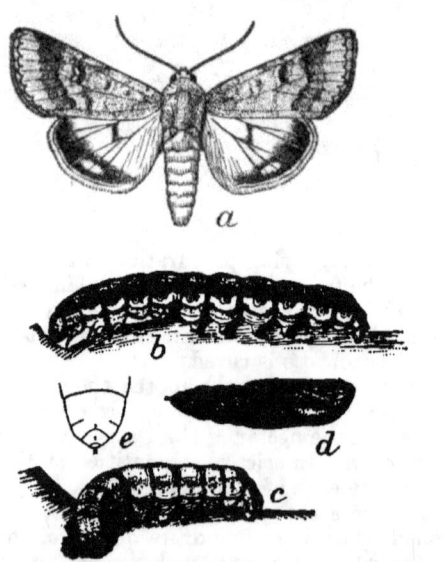

FIG. 8.—False bud worm or cotton boll worm (*Heliothis armiger*): *a*, adult moth; *b*, dark full-grown larva; *c*, light-coloured full-grown larva; *d*, pupa—natural size (original).

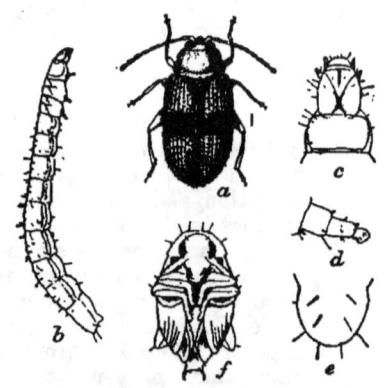

FIG. 1.—*Epitrix parvula*: *a*, adult beetle; *b*, larva, lateral view; *c*, head of larva; *d*, posterior leg of same; *e*, anal segment, dorsal view; *f*, pupa—*a*, *b*, *f* enlarged about fifteen times, *c*, *d*, *e* more enlarged (after Chittenden).

INSECTS INJURIOUS TO TOBACCO. (FROM REPORT OF U.S. DEPARTMENT OF AGRICULTURE.)

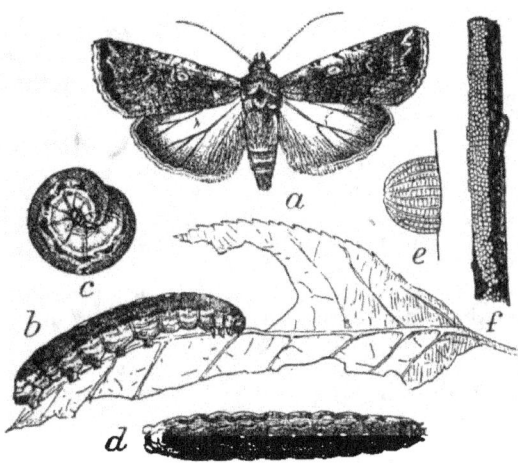

FIG. 16.—*Peridromia saucia:* a, adult; b, c, d, full-grown larvæ; e, f, eggs—all natural size except e, which is greatly enlarged (original).

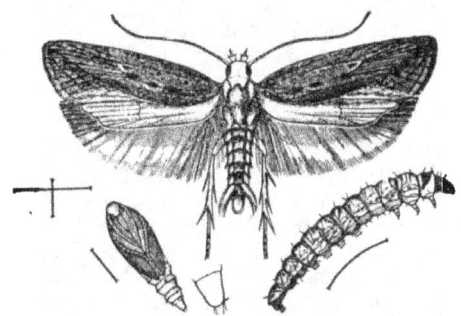

FIG. 14.—Tobacco split worm: adult moth above; larva below at right; pupa below at left, with side view of enlarged anal segment—all enlarged (original).

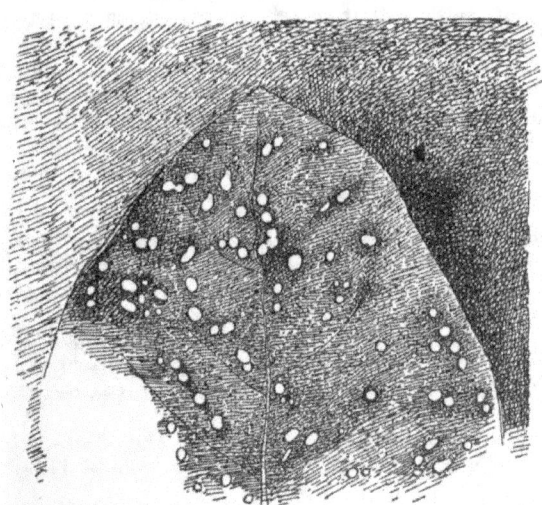

FIG. 3.—Leaf spots of old tobacco leaf—slightly reduced (original).

(FROM REPORT OF U.S. DEPARTMENT OF AGRICULTURE.)

At cutting time, care should be taken that none of these caterpillars be placed in the curing barn with the tobacco, for they will continue to eat the drying leaf as well as develop into parent moths for the next year's brood.

The Bud Caterpillars.—In America both the Heliothis Armiger and Heliothis Rhexiæ are known by the name of "bud worms," because of their habit of destroying the terminal bud of the plant. The Heliothis Armiger is very common in Rhodesia, and is the caterpillar that damages the mealie ears, the cotton, Cape gooseberry, and the tomato.

The spray of Paris green used for the horn caterpillars will also assist greatly in destroying this destructive pest, but if it becomes very common, it will be necessary to use the following special treatment. A pound of Paris green is mixed with fifty pounds of finely ground mealie meal, and a small quantity of the meal sprinkled on the bud of each plant. If the weather is at all wet, this treatment will have to be often repeated. This method is not as slow or expensive as it might appear to be, and is certainly very effective. In Florida where fine cigar leaf is grown, this sprinkling of poisoned meal is kept up throughout the season.

It is also a good plan to keep all the tomatoes and Cape gooseberries growing in the vicinity sprayed with Paris green. This will not only be a good thing for the plants sprayed, but will be of assistance to the tobacco, by destroying caterpillars that will produce later broods injurious to the tobacco plant.

The Leaf Miner or "Split Worm."—The larva of this moth, Gelechia Solanella, is already doing great damage to potatoes in Rhodesia, and will probably become a pest of tobacco. The moth, which is a very small gray one, lays its eggs upon the leaf, and the caterpillars as soon as hatched enter the leaf and there mine, between the two surfaces. Their presence is made known by a grey discolouration of this portion of the leaf. In potatoes these insects also work in the haulm and even in the stored tubers.

After working in one place for a time, the caterpillar will emerge on to the surface of the leaf, and enter at another place. Because of this fact it may be treated, to a certain extent, with arsenical poisons. The same spray that is used for the horn caterpillar is of value with this insect—that is, one pound of Paris green to one hundred and fifty to two hundred gallons of water. The potato fields should be as carefully treated as the tobacco, and it is very likely that this insect will also be found working on the tomato and Cape gooseberry, so that these plants should either be treated with poison or destroyed.

The Tobacco Flea Beetle.—This insect (*Epitrix parvula* is the common species of America) is a small active beetle that does great damage to all solanaceous plants—as the potato, tobacco, tomato, Cape gooseberry, datura, &c.

It eats the tobacco leaf full of very small holes, and thus destroys its value, as well as providing an entrance for fungus and bacterial diseases. A certain amount of damage is also done to the roots by the larvæ.

The preventive treatment, which is the destruction of all solanaceous weeds growing in the locality, should be adopted, for these weeds serve as breeding places for the earlier broods of this beetle.

When the beetle once starts its work on the crop, its ravages may be checked by the same Paris green spray that has been recommended for the other insects.

Lasioderma Serricorne, Fabr. A Beetle Injurious to Stored Tobacco.—This beetle feeds upon all forms of stored tobacco, and works injury by eating it full of holes. Cigars, cigarettes, and stored leaf of all kinds are equally enjoyed by this pest. Once it has taken possession of a curing barn, warehouse, factory or shop, it is there to stay, unless strenuous measures are adopted. In the warehouse the passing of the tobacco through the heat of the re-ordering machine will check the ravages somewhat, and the steaming of every portion of the room will serve as a temporary measure; but a few insects will be hidden in the cracks of the floor and wall, and serve to again infest the building.

Carbon di-sulphide has been recommended by Dr. Howard, of the United States Department of Agriculture, as a treatment for this beetle. The room is made as air-tight as possible, and the fumes of this gas are confined in it for at least twenty-four hours. The gas is explosive when confined, and is also deadly to all life, and for this reason no fire or persons are allowed in or near the building during the treatment, and the room is thoroughly aired before being again entered. In a tobacco shop all of the tobacco may be placed in a case and treated. One pound of the liquid for every thousand feet of space is placed in a vessel and permitted to evaporate in the room.

Other Tobacco Insects.—Other insects, as the grasshopper and the cricket, often do some damage to the tobacco crop. Any insect that injures the plant by eating can be destroyed by the use of arsenical poisons, as Paris green. However, if the insect secures its food by sucking, as do the true bugs (*Hemiptera*), it will be impossible to destroy with the arsenical poisons, and some spray will have to be applied that will kill the insect by contact. For the sucking insects the use of a concentrated solution of nicotine, diluted with sixty parts of water, has been found to be fairly effective.

All poisons should be labelled, and in their use care should be taken that they are not placed within the reach of young children.

Diseases of Growing Tobacco.

Mosaic Disease.—One of the most common diseases of growing tobacco is called "calico," or "mosaic" disease, because of the mosaic-like appearance of the light and green portions of the leaf. The disease causes the leaf to grow more rapidly near the veins than elsewhere, and thus become wrinkled and corrugated. A portion, or all of the plant, may be affected. Slightly diseased leaves are worthless as wrappers, and highly diseased leaves are of no value for any purpose. For many years the nature of this disease has been a mystery, and has been variously regarded as due to a fungus, as the result of an excess or deficiency of minerals in the soil, as produced by bacteria or induced by faulty drainage. Recently Dr. Woods, of the United States Department of Agriculture, who has carefully investigated the disease, has arrived at the conclusion that it is due to none of the generally supposed causes, but that it is "due to defective nutrition of the young dividing and rapidly growing cells, due to a lack of elaborated nitrogenous reserve food, accompanied by an abnormal increase in activity of oxidising enzymns in the diseased cells." These are the same enzymns that prove so beneficial in the fermentation process later. The enzymns are liberated by the decaying plants or roots, and, if in excess, may enter the roots of young plants set in the same soil, and induce a diseased condition, from which the plant will never completely recover. Plants with injured roots are more than commonly susceptible to attack.

The sowing of seed on fresh or burned plant beds, the avoidance of injury to roots in transplanting or in cultivation, and the making of conditions favourable to a steady, even growth, appear at present to be the only things within the reach of the planter for the prevention or moderation of this condition.

Seed from diseased plants should not be saved, for, while the disease may not be carried in this manner, still an inherited tendency to this condition may be transmitted.

"Frog Eye," or Leaf Spot.—This disease is also called "white speck," because of its appearance in the form of small white specks in the tissue of the leaf. It appears to a certain extent in nearly all tobaccos and does not do any large amount of damage. A few years ago cigar tobacco with this specking was in demand, but the style changed as soon as it was found possible to artificially produce this marking on any leaf. It is supposed by some to be caused by too much water at the tap root, and by others to be due to the presence of an excess of potash in the soil. It does not appear, however, that it is due to either of these causes, and is probably bacterial in its nature. This specking must be differentiated from the small white specks due to sun burning where there has been a particle of sand upon the leaf. No successful treatment is yet known.

Rust, or Blight.—There are two or three forms of what are known as rusts, or blights, which are physiological conditions caused either by excessively dry weather, or excessively wet weather, or by the use of too much of certain fertilizers. The farmers of South Carolina say that the use of large amounts of phosphates will cause the plants to prematurely ripen, which is sometimes called blight. The analysis of the conditions present in the field and the removal of the cause as far as possible is the only treatment.

There are several minor diseases of growing tobacco, but they can hardly be considered of interest unless they make their appearance in Rhodesia.

Diseases of Tobacco while Curing (Pole Burn, Pole Sweat, or House Burn).—Pole burn in the tobacco barn is due to excessive humidity, and is very likely to be present during prolonged warm wet weather. The disease is first noticed by the appearance of small dark spots near the stem and mid-rib. The spots rapidly increase in size and numbers, and in a short time become confluent. Within forty-eight hours the whole leaf, and, in fact, all the leaves in the curing barn, may be affected and destroyed. The tobacco becomes very dark in colour and thoroughly decayed.

This decay is due to a bacterium that gains entrance to the leaf through any broken place or through openings made by fungus growths. Temperatures of 110° stop its action, as also does the reduction of the humidity of the room. For this reason the disease may be controlled through regulation of the temperature and humidity of the room by means of fire and ventilation. The maintenance of a stove in the curing barn is always to be recommended. If there is no stove, small open charcoal fires may be started. All decaying leaves should at once be taken out of the barn and destroyed, to prevent infection of the remaining tobacco.

During wet weather this disease should be watched for. It will probably first make its appearance in the centre, or in the least ventilated portion of the room.

Stem Rot.—This is due to a fungus, Botrytis longibranchiata. It appears in the form of a velvety white mould on the stalks and stems of the curing tobacco, and is most prevalent during wet weather. If it has once made its appearance in the curing barn further trouble should be prevented by thoroughly cleaning and disinfecting the barn before again using. This may be done by sweeping the walls and floor of the barn, and then washing with a mixture of lime and sulphur in water, which, if made sufficiently fine, may be sprayed on with an ordinary spray pump. The burning of sulphur in the closed barn will also help to destroy the spores of the fungus. Heat will serve as a check in unfavourable weather.

White Veins.—These may make their appearance in the barn after a long spell of dry weather. So far as is known they are due to the fact that the outer cells of the leaf are killed too soon by the rapid drying of the leaf, and allow the admission of air under the epidermis, thus giving the appearance of an absence of colouring matter in the leaf. This condition injures the value of the leaf for wrappers, but seldom for any other purpose. The only way suggested for the prevention of this condition is to keep the air of the barn slightly humid in dry weather by sprinkling water on the floor.

Moulds and Rots in Cured Tobacco.—After tobacco has been cured, different moulds and rots often do much damage to the stored tobacco. These will greatly injure or absolutely ruin the leaf. The development of these troubles is due to the presence of too large a quantity of moisture in the leaf. The tobacco should be stored in as dry a condition as possible and examined now and again to see that it does not absorb moisture from the air. A stove in the room to heat the air occasionally when the weather is wet will largely prevent the attacks. A room where the tobacco has once moulded should be thoroughly cleansed and disinfected before tobacco is again stored in it.

In the case of the black rot, cigar fillers may be put through what is known as a forced sweat, to kill the fungi and drive away the musty odour. The tobacco is allowed a large quantity of water, and placed in a warm room where it will heat up rapidly, and in four or five days, in warm weather, will have reached as high a temperature as is safe. This sweat kills the fungi and if carried out thoroughly will also destroy the spores. The tobacco is not likely to be of the highest quality.

"**Saltpetre**"—The mould-like appearance that often comes over tobacco while curing and fermenting is called "saltpetre" and is due to a saline efflorescence caused by the presence of large quantities of salts in the leaf, as potassium, sodium, calcium, and magnesia. A light brushing and a spray of a four per cent solution of acetic acid will remove this for the time.

DAMAGE BY WIND AND HAIL STORMS.

Heavy wind storms often break off the upper leaves of the plant and whip the lower leaves around on the ground until they are badly torn. The planting of tobacco in sheltered fields is the only measure that can be adopted against damage by the wind. Where there are no natural windbreaks the planting of large numbers of rapid growing trees is to be recommended. A temporary expedient is to plant several rows of mealies (Indian maize) on the windward side of the field. Strips of mealies may also be sown at intervals of six or eight rods throughout the tobacco field. If leaves are accidentally broken off the tobacco plant when it is nearly ready to harvest, they may be gathered up and cured.

Even a very light hailstorm will injure the value of a leaf for wrapper purposes, and a very heavy hailstorm will entirely destroy the whole plant. Where a plant has been badly broken it may be cut off near the ground and a sucker left to develop a new plant.

In some localities in Europe a method of breaking up hailstorms and preventing damage to the vineyards is practised. It consists in the simultaneous discharge of many cannon that have previously been placed at different points in the district. When this practice has been fully tested and its merits more fully determined it may be feasible to undertake it in the more thickly settled tobacco districts.

Throughout America certain companies will, for a premium of about four per cent, insure the tobacco crop against damage by hail or wind. The amount of the damage is usually determined by a committee of non-interested tobacco men, who are appointed for the special case and are remunerated for their services by the insurance company.

CURING.

Curing is not merely drying, but is a chemical process the exact changes and reactions of which are not fully understood. The quality of a tobacco is made in its growth; curing but fixes or further develops those qualities. A badly grown tobacco cannot be made into a high-class product by any process of curing, although by skill in handling it may have latent or slightly developed good qualities brought out and emphasized, and it may also have its bad points partially suppressed. A very fine tobacco may be absolutely ruined by lack of skill in the curing process.

The process may be said to commence the moment the plant is cut and to continue until no further change takes place in the leaf.

Commonly this curing process is divided into several stages, the first one of which is allowed to appropriate the word, "curing." The second stage is known as the "fermentation" or "sweat," and there may also be a third stage which is known as "ageing." This third stage is but a mild continuation of the process of fermentation.

The theory that bacteria are largely instrumental in producing the many changes of the different stages of the curing process, has been advanced by a number of scientists. It was even, for a time, supposed that bacteria produced on the aromatic Vuelta Abajo could be transferred to other tobacco, and that the fermentative processes thereby initiated would develop an aroma equal to that of the Cuban tobacco. From this it was reasoned that all distinctions of section, soil, and climate would be broken down, and all that would be necessary for the production of fine tobacco would be to select a locality where the soil was fertile and labour cheap, and inoculate the tobacco with the best aroma-producing bacteria. However, no startling commercial changes based on this theory have as yet taken place.

Very recently Dr. Loew, of the United States Department of Agriculture, has shown that not only are bacteria not responsible for the fermentation of tobacco, but that the fermenting leaves are destructive of bacterial life. By a series of experiments, he has demonstrated that the chemical changes that take place in the curing and fermentation of tobacco are due to the presence of oxidizing enzymns. Enzymns are closely related to the soluble ferments. One of these ferments, diastase, takes a prominent part in the fermentation of malt, and will change two thousand parts of starch into sugar for each part of itself. The oxidizing ferments have the power of taking oxygen from the air, and supplying it to the contents of the plant cells, thus causing the splitting up of existing chemical forms, and the creation of new products. In this process the enzymns suffer but little loss of themselves, for they merely act as agents, and take with one hand what they give with the other. Platinum black has a somewhat similar power, as is often shown in the chemical laboratory. An example has been given of a somewhat analagous action that takes place when a weather-exposed board decays more rapidly in the proximity of a rusty nail. In this case the wood, assisted by the iron oxide, is enabled to combine with the oxygen more rapidly than it would if left to itself.

Enzymns are highly complex protein forms, and make up a part of the protoplasm of the plant. They are easily destroyed or changed into other protein forms by much heat, or by the too rapid loss of their moisture. When the plant is slowly starving to death, as it is when it is cut and allowed slowly to dry, there is a rapid formation of these enzymns, which separate themselves from the protoplasm, and push out through the plant in search of food for the dying plant cells. Having thus distributed themselves, the enzymns are in position to become again soluble, and take up the work of fermentation whenever the conditions become favourable, as they do in the fermentation pile. If the leaf be killed by heat or by rapid drying, the enzymns will have no opportunity to escape from the protoplasm, but will become entangled with the insoluble protein, so that later, when the leaf is moistened for the fermentation, the process will be a partial or total failure, for these enzymns are only active when in solution.

To repeat, the chemical changes in the leaf which develop the aroma, as well as eliminate undesirable products, are due to certain enzymns. These changes take place during the second stage of the curing process, which is commonly called the fermentation or the sweat. But these enzymns are largely developed in the first stage of curing, and unless this first stage be properly conducted, the enzymns will not exist in large quantities, or in available forms, and the products developed in the fermentation will be largely disappointing. With this fact clearly in mind, the reasons for certain steps in the curing will be better understood.

Fine aroma is considered of greater importance in cigar leaf than in other forms of tobacco, and for that reason greater attention is given to its fermentation, and more care is exercised in its preliminary curing, so as to prepare it for fermentation. All tobaccos go through a fermentation, whether such a stage be regarded as a part of the routine or not. In some tobaccos, where other characteristics than aroma are chiefly sought, the curing process may be such as to destroy the enzymns and the leaf's power of fermentation. This is largely the case with the Bright yellow tobaccos, for the intense heat and rapid curing of the process designed for the colour, destroys most of the enzymns. But even in this case, a small amount of the enzymns must survive, for the cured leaf goes through a slight fermentation known as the "May sweat." It is in this sweat that most of the aroma that this type of tobacco has is developed. If this tobacco be placed in a moist condition and bulked, as are the cigar tobaccos, it will not ferment, but will decay. The tobacco grower will say that the texture of the leaf has been destroyed by the heat; it would be more accurate for him to say that the oxidizing enzymns have been destroyed.

Several different methods of curing are in vogue. The different methods are based on differences in the nature of the tobacco grown, as well as on fuel supply, labour conditions, and the demands of the market. Sometimes the processes do not appear to be based on any modern conditions, but to be merely survivals of earlier times. In the past, great carelessness was the rule in all methods, but now with a better understanding of the processes involved, and with the requirements of a constantly more exacting market to cater for, a great change is taking place for the better. Different types of curing barns are in use for the different methods of curing, but it is seldom that any of these barns have as yet reached a state approximating perfection. Many of them have been built with the idea of utilizing the material at the builder's disposal, rather than with any thought of their adaptability to the purpose for which constructed. This matter will be discussed in the chapter on buildings.

The terms describing the different methods of curing are based on the most important feature of the method. Thus we have sun curing, fire curing, flue curing and air curing.

Sun Curing.—This is a method but little used, and is confined to a small section of Virginia, where the production of a very sweet chewing tobacco is the feature sought after. The tobacco is hung on racks in the sun, care being taken that it is shaded enough, during the wilting period, to prevent sun-burning. These racks are placed near a building, so that the tobacco may be carried under shelter upon the approach of a storm. This additional handling of the tobacco, whenever the weather conditions are unfavourable, involves a large amount of labour, so that the acreage grown is necessarily limited. The presence of much sugar in this tobacco renders it very liable to the attacks of moulds, so that, unless careful attention is given to the handling and storing of this sun-cured tobacco, there will be heavy losses. The room where this tobacco is stored should be fitted with a stove so that the tobacco may be kept dry in wet weather. Little money is made out of this class of tobacco in America, and it is largely grown by poor farmers whose families assist them in the handling of the crop.

Fire Curing.—This is a system whereby the tobacco is hung in barns and cured by means of open wood fires, the heat and smoke of which both have an important part in the process.

The greater portion of the tobacco shipped to Europe and Africa is cured by this method. In the early times, tobacco that had been heavily smoked was able to withstand the ocean voyage far better than tobacco that had been cured in the air. In time, however, the European palates and olfactories have become so accustomed to its creosotic flavour and "ham" odour that it is chosen in preference to the better tobaccos. At the time of writing the demand seems to be changing from this "fired" tobacco to the better types, and it appears probable that in time this tobacco will largely disappear from the markets.

The tobacco is harvested with the stalk, and after being allowed to wilt for a short time is hauled to the barn, placed astride the curing sticks at the rate of from six to eight plants to each four foot stick, and hung on the rafters or poles. The tobacco is allowed to hang until it becomes a rich yellow colour; to reach this stage, will, probably, require four or five days. This yellow colour is owing to a chemical change in the chlorophyl of the plant, due to absence of sunlight. When this stage is reached, small slow fires are started on the floor under the hanging plants. The heat is not permitted to rise above 90° for at least twelve hours, and then is gradually increased until it reaches 150° in four or five days, when the tobacco should be dry and the fire may be allowed to burn out. Even then the stem will be full of moisture, and, as soon as the heat is lowered, the water of the stem will spread out through the fibre of the leaf and make it soft. A new fire is then built to again dry the leaf, and this is repeated whenever the leaf shows a tendency to become soft. This second drying is particularly required if the lighter shades are desired in the leaf. It should be understood that this curing process must be gradual, for if the heat is allowed to rise too soon, or rapidly, it will cook the leaf and give it a bluish tinge. If the fires are started before the leaf has become sufficiently yellow, the leaf will be stiff, and have but little flexibility. For the English markets the tobacco is cut before fully ripe, resulting in an olive green colour. If firing be delayed too long the tobacco may suffer from "pole burn." There will also be losses from this trouble, if the leaf be allowed to hang in a very moist condition in the barn during or after curing.

As soon as the tobacco is cured it is stripped from the stalk and graded. It is then tied into "hands" and piled in bulks, which are covered with canvas.

In tying into "hands," anything from six to twenty leaves are taken in the left hand, the stem ends are placed in the same direction and evenly together; a small leaf is then selected, smoothed out at the tail and doubled over so as to make a band an inch wide, and this leaf is wrapped tightly around the stems of the other leaves, and has its stem end tucked between and through them. Upon the evenness and smoothness of this tying depends somewhat the selling price of the tobacco. Some operators pass each bundle through the hand as tied, so as to press the leaves together and give a better appearance.

No attempt is made to carry on fermentation in these bulks. Some changes do take place, however, and it is necessary to occasionally observe the pile and not allow the temperature to become too high. When the tobacco is in the right condition, and contains enough moisture to permit of its being handled without breaking, it is carried to the market and sold in the loose condition, or is packed in hogsheads and sold when markets are favourable.

Where fuel is scarce the item of wood makes a large reduction in the profits. The wood must be dry and well seasoned, for a fuel that gives off soot, creates a heavy smoke, or has a disagreeable odour, will destroy the value of the tobacco. The fires are made by keeping the ends of several sticks together, as the Kaffirs do. Care must be taken not to let the blaze rise high, for the drying tobacco will easily catch fire and destroy itself as well as the barn.

Flue Curing.—This also is a process where artificial heat is used for curing the tobacco, but in this case open fires are not permitted and the smoke does not come in contact with the tobacco. The fires which are of wood (in Japan coal has been used), are in small brick furnaces on the outside of the building, and the heat is carried through the building and under the tobacco by means of large sheet iron pipes or flues.

This is the system used for the curing of the yellow tobacco, which has become so popular for pipe smoking, cigarettes and chewing, particularly in the last case for plug wrappers. The feature sought in the yellow tobacco is the colour, and the aim is to produce this, and yet damage the texture and elasticity of the wrapper leaves as little as possible.

No other type of tobacco or other system of curing requires as much skill in its handling. A little misjudgment in maintaining a certain temperature for too long or too short a time will largely lessen the value of the product. Slight shades in colour may mean large differences in the selling price. A leaf that might become a wrapper, if injured for that purpose, will have its value reduced 80 per cent. No set rules can be given for the handling of this process, as much depends on the condition of the leaf when placed in the barn and on the weather conditions during the time of curing.

The barns will be described later, but it may be said here that the barn must not be so large that it cannot be filled in one day, for the tobacco in any one barn should not be in different stages of greenness. The barn must also be in a locality protected from the winds, for a strong wind will cause the temperature to be much lower on the windward side of the barn than elsewhere, and thus prevent an even curing of the contents. The ventilation of the barn must also be under perfect control. It is perhaps needless to say that, where so much heat is used, a thermometer is a necessity. This should be hung in the centre of the barn, and in such a position that it may be easily read from the door.

The tobacco may be cured on the stalk, or may be cured by the single leaf system. If the single leaf system is used, more pounds may be placed in the barn, and each barn's contents will be largely of the same degree of ripeness and of the same grade. The stalk system will require more time in the curing, but less time in the harvesting. The sap of the stalk is also, to a certain extent, translocated to the leaf, giving the leaf greater weight and gumminess.

Whether the leaf or the stalk method be adopted, the same four foot curing sticks are used.

As soon as the barn is filled the fire is started, and from that time until the barn is finished, at the end of four or five days, it is not allowed to die down or go out. The curing process may be divided into three stages.

The first stage is known as the "yellowing" or "wilting" stage. The heat is kept at 90° for from eighteen to twenty hours, or until the tobacco has reached the proper colour and lost a portion of its moisture. Some curers hasten this process by holding the temperature at 90° for but three hours and then rapidly increasing to 120°, and as soon as that point is reached dropping back to 90° for another six or eight hours. During this stage certain life chemical changes take place in the leaf; there is a destruction of chlorophyll and starch, and the creation of enzymns.

The second stage is for the purpose of "fixing" the colour. The yellow colour is already in the leaf, but would, by oxidation of the cell contents, change to a brown or red, unless it be fixed by the application of heat and the exhaustion of moisture. This loss of moisture and the application of heat destroy the oxidising enzymns that are responsible for the change in the cell contents, and the brown colour. To destroy this enzymn, and fix the colour, requires a temperature starting at 100° and gradually rising to 120° during a period of about twenty hours.

The next stage is what is known as "killing" the leaf by the final exhausting of the moisture. The temperature is held for some time at 120°, and then gradually rises to as high as 135° and even 140°. This process takes about forty-eight hours, at the end of which time the leaf is so dry that it will crumble to powder in the hand. If the single leaf method is used, this high temperature must be maintained, or even increased, until the stem is perfectly dry and brittle.

If the tobacco is being cured on the stalk, a fourth stage is necessary, and is known as "killing the stalk." The temperature is gradually increased at a regular rate per hour, until it reaches as high as 160° or 175°, and the stalk is thoroughly cured out. If the stalk be not thoroughly cured as soon as the temperature is lowered, its sap will spread into the leaves and form red places in the region of the veins.

Many different modifications of this flue-curing system are in use, differing from each other largely in the degree of temperature allowed at different stages. In fact, no set rules can be given, and the formula may be varied for each barn cured. This is a matter for the judgment of the man in charge, and to give a large number of different methods would only confuse the reader. This method of curing cannot well be learned from a written description.

The leaf must be forced to sweat, and will do so at a temperature of from 110° to 120°. Whenever the sweat commences, the temperature must be maintained until it is completed. This sweat will carry out of the plant many disagreeable substances that would be harmful to the quality of the tobacco. As soon as the leaf indicates that the sweating is at an end, by drying off, the fire must be lowered, the ventilators opened, and the barn cooled off. Sometimes the temperature is run up to 120°, and then lowered to 105°, by ventilation, several times in succession. If the barn be so dry that the sweat will not commence, assistance may be rendered by moistening the air by means of pans of water or the placing of wet grass on the floor.

If the temperature of the barn be forced too rapidly the sap will be dried into the leaf, and this will be shown by the presence of red blotches or spots. If the temperature be not advanced rapidly enough the leaf will "sponge." By this is meant the presence of brown porous places on the surface of the leaf. If the temperature is advancing too rapidly the fact will be indicated by the browning and curling of the edges and "tails" of the leaf. A high temperature too early in the process will cook or scald the leaf. Often the tails of the lower tier of leaves in the barn will be scalded, and for this reason it is best to place the worst leaves on the bottom.

No sticks or leaves must be allowed to fall against the flues, for in the last stages of the curing the tobacco will easily catch fire.

About three cords of dry wood are required for each thousand pounds of tobacco cured, and this means a considerable item of expenditure in localities where fuel is expensive. Soft coal may be substituted, and with it the fire is more easily controlled than with the wood.

As soon as the tobacco is cured the fires are extinguished, and the doors and ventilators opened. The moisture of the air will soon bring the leaf in "order" or condition for handling. If the weather be wet the barn should not be opened, for it would damage the tobacco, in that it would become too moist. If the cured leaf does become too moist it must be again dried with a new fire. If the air be very dry the moisture content of the barn may be increased, and the "ordering" process hastened by the use of water on the barn floor, or the placing of moist grass upon the flues.

As soon as the tobacco is in condition, it is taken down and bulked without being removed from the sticks. A few days of bulking will straighten out the leaf and improve its appearance. Next the tobacco is hung in the packing house. It is hung very closely so as to be influenced by weather conditions as little as possible. Later in the season it is taken down, graded, tied into "hands," and again re-hung or bulked until the grower wishes to market it. Care is always taken that it does not become in too moist condition for this would darken the colour and reduce the

CURED TOBACCO, STRIPPED AND TIED IN HANDS AND RE-HUNG IN THE CURING BARN.

value, as well as render the leaf liable to mould. Care should be exercised not to handle the leaf when it is so dry that it will break, for the most valuable leaves are those intended for plug wrappers, which must not be injured in any way. Many growers have a cellar under their packing shed or at the edge of the barn, where the tobacco is hung to become moist and pliable enough to handle.

This tobacco is sold loose to the re-handlers and manufacturers, and may be marketed at any time after it has come out of the barn. However, its quality and value certainly increase with age, and two-year-old tobacco commands a premium.

A greenish tinge that may show in some of the leaves will largely or totally disappear as the tobacco ages.

Air Curing.—All the different types of cigar leaf as well as the White Burley tobacco are air cured. Any tobacco may be air cured, and this is the method that should be adopted in most of the tobacco-growing regions. Of course, where the yellow colour is the characteristic sought after, heat must be applied, and for markets desiring a smoky flavour there must be some smoke; but the method of using the air for curing, and of using artificial heat only in times of unfavourable weather, is the one that is likely to be generally adopted.

The White Burley is cured on the stalk. It is wilted and hung in the barn in the same manner as other stalk-cured tobacco. Sometimes the tobacco is first hung on scaffolds in the field, and left there to wilt for several days before being taken to the barn, but if this is done the tobacco may suffer from unfavourable weather conditions. If the barn is properly constructed and can be easily ventilated, the plants may be at once hung in the barn. The process of curing lasts for about six weeks, and consists in regulating the temperature and humidity of the air by means of the ventilators. If the leaf is drying too rapidly the ventilators are opened on moist days and nights and closed on dry ones, or if the tobacco is slow in the process, and is likely to be attacked by fungus or bacterial disease, the leaf is dried out by having the ventilators opened on dry days. If there is a period of continual wet weather, the barn may be dried somewhat by the use of charcoal fires, or a stove may be set up in the barn and a fire started.

When the tobacco has been thoroughly cured, which can be determined by the condition of the stems and stalks, it may be taken down and stripped. In stripping tobacco, the butt of the stalk is held in the left hand and the leaves pulled off one by one. The leaves, as soon as stripped from the stalk, are graded and tied into "hands." The tobacco is then bulked until it is ready to place in hogsheads. If the tobacco has a large moisture content when bulked the pile will ferment, and must be broken up and re-made.

The White Burley has a large absorptive capacity, and is in demand by manufacturers of chewing tobacco. The lighter grades are also used for pipes and cigarettes.

Cigar Leaf.—Most types of cigar leaf are cured on the stalk; but in certain types, as the Sumatra, when the plant ripens unevenly, and where a slight difference in the ripeness of the leaf intended for wrappers makes a large difference in the value, it is necessary to harvest the crop by the single leaf method. In this case the leaves are strung on twine with a needle, and placed back to back, so that leaves cannot fold round their mates and prevent proper drying. When the entire plant is harvested, the stalks are speared on the curing stick.

The after fermentation is of great importance to cigar leaf, and the curing process must be conducted with the idea of developing the enzymns as much as possible. As long as the cells of the plant are alive (and if properly handled they will remain alive for several weeks), there is a movement of organic matter from the cells of the tissue to those of the ribs; thus the starch in the form of sugar is largely transported and consumed by the processes of respiration.

In curing cigar leaf, the attempt is made to have the leaf become dry and come into condition (moist) once in twenty-four hours. More of the oxidizing enzymns are formed in the ribs of the leaf than in the tissue, and this alternate drying and moistening of the leaf brings about a movement of the contents of the ribs out into the body of the leaf. The longer the ribs and the stem of the leaf are permitted to live, the greater will be the amount of enzymns formed.

Whenever pole-burn or rot is detected, the barn may be dried out by means of charcoal fires or the use of a stove. If the climate be an extremely dry one, so that the leaf will not come into a moist condition at least once a day, the air of the barn must be rendered humid by the use of water on the floor. Steam may also be turned into the barn for this purpose. The ideal tobacco barn of the future will probably be fitted with steam heat, so that the moisture and temperature conditions will be absolutely under control.

When cured, the tobacco may be allowed to hang in the barn until a less busy season, or it may be taken down, stripped, sorted and bulked for fermentation. In cold weather tobacco leaf may appear to be thoroughly dry, but as soon as the temperature rises the tobacco will become soft and moist. For this reason if tobacco be taken down from the curing barn and bulked on a cold day it will contain more moisture than is apparent, and will rot or mould when the weather becomes warmer.

If cigar leaf be cured out too rapidly the leaf will be stiff and woody, and the colour will be uneven. If the curing be protracted too long a time, the leaf will lose its elasticity and strength.

In dry weather the barn should be closed in the day-time and opened at night. This will give the required moisture conditions.

There is this general difference between leaf cured by natural methods and that cured by artificial methods. The air-cured leaf preserves its flavour and is free from smoky odours. It also has a large absorptive capacity, because its porous system has been left free and open. The fire-cured leaf has less absorptive capacity, for the reason that the heat has contracted the porous system and the smoke has filled the surface of the leaf. The smoke preserves the leaf and enables it better to withstand exposure and unfavourable environment. The same tobacco when cured by fire brings on an average a penny a pound less than it would have done if cured by air. Very light charcoal or closed fires, such as are used to dry out a building in unfavourable weather, will not injure the quality in any way.

A LARGE FERMENTATION PILE, OHIO.

Fermentation of Cigar Leaf.—The fermentation, or sweat, is for the purpose of developing the aroma of the tobacco. On the ability of the leaf to properly ferment, and on the skill of the man in charge to regulate that fermentation, depend the quality of the finished product. This fermentation, as before stated, is due to the action of enzymns, and these enzymns must be subjected to certain heat and moisture conditions before they can begin their action. Tobacco should contain 23 per cent. to 24 per cent. of moisture to insure a proper fermentation; tobacco with a higher percentage of moisture is more subject to decay, and tobacco with less moisture will not ferment, or if fermentation does commence there will not be moisture enough present for its completion. The proper condition is soon learned by experience, and no test except the feeling to the hand and the pliability of the leaf is required.

The fermentation process is not usually conducted by the grower himself, but is handled by dealers who ferment hundreds of thousands of pounds in a season. The planter is not ordinarily in a position to give his full attention to this work, nor has he buildings where the proper heat and moisture conditions can be maintained. There is no reason, however, why the very large planter should not combine the functions of a grower and leaf handler. This is done by the tobacco growing companies in Florida.

Several different methods of fermentation are practised. In the method that has been most largely used, until recently, the tobacco is packed in wooden cases holding about three hundred pounds of tobacco to a case. The butts of the "hands" are placed to the outside and the tips to the centre of the case. By means of a screw or lever the tobacco is pressed down moderately tight and as much air as possible excluded. The top of the box is then screwed on and the case placed in a room that is kept at an even temperature. The box has moderate sized openings between the boards, so as to allow for the escape of the moisture and other waste products of fermentation. The tobacco is left in these cases for a summer and is then sampled and repacked to await sale to the manufacturer. At times the cases may be placed in heated rooms and the fermentation forced. This method of fermentation does not give altogether satisfactory results. The tobacco in the cases cannot be observed and there is no knowledge as to whether the fermentation is proceeding properly or not. To-day the tendency among progressive tobacco men is towards the adoption of what is known as bulk fermentation.

Bulk Fermentation.—The best rooms for this purpose are heated with steam and kept at a temperature of from 75 degrees to 80 degrees, and the humidity maintained at 80 degrees to 90 degrees, and even as high at times as 100 degrees.

When the tobacco is received for fermentation it is sorted into three different grades, according to the colour and texture. This is done so that the different grades may be fermented separately and given different treatment according to the nature of the finished product desired. Just at present the demand is for light shades in the wrappers, and if the wrapper leaves be given as heavy a fermentation as the filler leaves their colour will become dark.

When graded the tobaccos are placed in separate bulks. The number of pounds of the lighter grades allowed to the bulk is from three to five thousand; of the medium grades from eight to ten

BUILDING BULKS OF CIGAR LEAF FOR FERMENTATION, OHIO.

thousand; and of the ordinary fillers from ten to thirty thousand. The greater the fermentation desired the greater the percentage of moisture that the leaf is allowed to retain when bulked. The wrappers, therefore, are in a somewhat dryer condition than the fillers. These bulks may be from four to five feet wide, from four to eight feet high, and of any length. The length and the number of pounds placed in a bulk is limited somewhat by the labour at the disposal of the fermenter, for often, when the temperature is rising rapidly, the bulk must be taken down and rebuilt in a very short time, and it takes a considerable number of persons to properly handle, say, thirty thousand pounds of cigar leaf. No pressure beyond the weight of the tobacco is applied to the bulk, for it is desired that there be some space for the movement of air and the escape of the products of fermentation. The bulk is not built directly on the floor, but is placed on a platform raised a few inches above the floor. This platform is covered with a layer of wrapping paper. The butts of the tobacco are placed toward the outside of the bulk and the tips toward the centre. The first row is laid with the butts even with the edge of the bulk, the second row is placed so that the butts rest on about one-third of the tip end of the first row, and so on with the third row. Three rows in from each side, or six rows in all, is all that the ordinary bulk will require. The process is repeated until the bulk has reached the desired height. Where fine wrapper leaf is being bulked, strips of wrapping paper are often placed under the butts of each row so as to prevent injury to the leaf under it. When the bulk is completed it is covered with canvas, blankets, or rubber sheeting.

The temperature will begin to rise in a short time and will continue to increase at the rate of from five to fifteen degrees per day, depending on the percentage of moisture present, until the temperature of the pile reaches 130°, when the bulk must be broken down and rebuilt.

In building the new bulk the tops and sides of the old bulk should form the centre of the new. Each "hand" should be given a shake as passed over to free it from any of the objectional products of fermentation, and lessen liability to rot and mould. The temperature of the tobacco will be lowered in handling to about the temperature of the room. The bulk will again heat up, but not so rapidly perhaps as the first bulk. In eight to twelve days the thermometer will indicate that the pile has reached a heat of 125° or 130°, or that perhaps it has ceased to rise in temperature and remains stationary. In either case the bulk is to be rebuilt.

This process of rebulking may have to be repeated six or eight times, or until the best aroma possible is obtained. If the process be carried too far, the desirable products obtained in the earlier fermentation may be destroyed and the tobacco left about as valuable for smoking purposes as old rags or paper. The lighter coloured wrappers must not be heated as highly, or fermented as long as they would stand, or as much as might be desirable from the standpoint of aroma, for their chief value lies in their colour, and this must be preserved.

If the tobacco be too moist in the bulk, as will be indicated by its "sogginess" or a very rapid rise in temperature, it must be rebulked more often than otherwise and in the rebulking should be handled in a dry room, which will deprive it of a portion of its moisture.

If the tobacco be found to be so dry in the bulk that fermentation ceases or is retarded, this defect may be corrected by handling it in a

88 THE CULTURE OF TOBACCO.

MAKING CUBAN CARROTTES (U.S. DEPT. OF AG.).

warm, moist, or steam-charged room. In some districts the use of steam is objected to, on the ground that it may at times give to tobacco an objectionable odour. In these places compressed air is made to create a fine spray of water and thus moisten the air of the room and serve the same purpose as the steam. For the fillers, if deficient in moisture, a dipping of the butts in a cask of water will be sufficient unless they are lacking in gum, in which case it is a common practice to dip the butts in a preparation made by boiling a quantity of Havana tobacco stems, and mixing the resulting thick juice with sour wine at the rate of three to one. After being dipped in this, the tobacco is placed in cases or small piles and covered up for a day, so that the moisture may become evenly distributed. The new bulk is then made. A fine mist of steam or spray of water may be added to the tobacco, but the direct addition of water may slightly injure the colour of the leaf.

When low grade fillers cease fermentation before the desired stage is reached they are treated with the following "petuning" solution:—Two gallons rum; one gallon sour wine; one half-pint tincture of valerian; one ounce oil of aniseed; one half-gallon black coffee; one ounce pulverized cloves; one ounce pulverized cinnamon; two pounds liquorice paste dissolved in water and sufficient water added to make five gallons. After being allowed to stand for twenty-four hours and thoroughly mixed, the preparation is ready for use. As the bulk of the tobacco is being made a fine spray of this is placed on each layer. The moisture added aids somewhat in the process of fermentation, but the main idea of this preparation is to add to the tobacco an artificial aroma resembling that of Cuban tobacco. This "petuning" is never done to high-grade tobacco.

When a tobacco is slow in heating it is sometimes sprayed with a solution of ammonia carbonate. The reason for this is that the contents of the leaf give an acid reaction because of the accumulation of free acids, and the ammonia carbonate combines with these acids and gives a neutral condition favourable to the action of oxidizing enzymns.

A temperature of 130° is probably as high as the tobacco should ever be allowed to rise. Expert tobacco men will judge the condition of a pile by the insertion of their arm, but most persons should trust nothing but an accurate thermometer. This may be read at any time by keeping it in a perforated tin cylinder inserted into the centre of the bulk. The United States Department of Agriculture has recently devised an electrical thermometer which may be left in the centre of the bulk, and the readings made with no disturbance of the tobacco. The wires may be extended to a central office and the readings of any number of bulks in different buildings be observed there.

Upon the completion of the fermentation the tobacco is ready to be graded and packed for shipment. Tobacco is in a proper condition for baling when a "hand" that has been squeezed may have all of its leaves separated one from another by a shake. This indicates that the leaves are without enough moisture to cause them to stick one to another. The finer grades of cigar leaf are graded very closely and handled very carefully. The commoner grades of fillers are placed in three-hundred pound cases. The best grades of Cuban tobacco are made into "hands" of forty leaves each, and then four "hands" are bound together by means of Cuban bast into what is known

CARROTTES OF FLORIDA CUBAN TOBACCO (U.S. DEPT. OF AG.).

A SUMATRA BALING PRESS AND A BALE

CIGAR LEAF IN CUBAN BALES.

as a "carrotte." Each carrotte weighs a pound or a little over. Great care is taken that the outside, or wrapper leaves, of the carrotte be smooth and presentable. Eighty carrottes are then made into a bale, which is pressed into shape in a press and is covered first with canvas and then with the inside bark of a Cuban tree, Only the best grades of goods are made into carrottes. The second grade, or damaged leaf, is given a thorough second sweat and stemmed, smoothed, and flattened out, and made into what is known as a " book of fillers." In stemming, not all of the mid-rib is removed, but a piece about two inches long is allowed to remain in the tip of the leaf. These books are then made into Cuban bales.

The Sumatra tobacco is made into " hands " of about forty leaves, each leaf being folded. These hands are packed into bales, being spread out fan-shaped and carefully flattened as packed. The fillers are made into Cuban " carrottes " and the broken leaves into " books." The bales are flat and are covered with matting, as is done in Sumatra.

Each bale or package of tobacco should be carefully labelled as to grade, weight, time of packing, and the name of the seller, so that any errors may be traced out and corrected and the reputation of the packer established.

The bales of the best wrappers, where no fermentation is desired, are placed in a cool room. If further fermentation is wished, as in the case of the fillers, they are placed in a warm room. The wrapper bales should be stood on end and reversed every other day for several weeks. They may then be piled one upon another. The filler bales may be piled at once and their position reversed at least once a week. The wrappers will be in condition to manufacture in three months after baling, and the fillers in six months.

Ageing.—After a tobacco has been cured, it must go through a process of ageing before it is fit for consumption. Fermented tobaccos require less ageing than the ordinary unfermented tobaccos. Smoking tobaccos are allowed to age for at least two years before being manufactured, and often the process is continued for four or five years. After five years there is likely to be a slight loss rather than any improvement in quality. Ageing may be described as partially a process of slow fermentation and partially an oxidation of the leaf contents without the agency of enzymns. Ageing certainly softens and mellows a tobacco, taking away its rawness and bitterness as well as disagreeable odours, and improving the aroma as well as the burning qualities of the leaf. No tobacco, and particularly no tobacco from a district that is trying to obtain recognition, should be marketed before it has developed the qualities that come with age.

SOME OF THE CHEMISTRY OF CURING AND FERMENTATION.

During the fermentation of tobacco there is a loss of as high as fifteen per cent. in weight. Part of this is due to the loss of moisture and a part to the loss of solid matter through the decomposition of different products and the development of gases. The presence of ammonia is easily detected by the odour in the fermentation room. This is produced as a result of chemical changes in the tobacco.

The starch largely changes to sugar, and the sugar formed is largely consumed in the curing process; the remainder of both of these products is usually destroyed in the fermentation.

As soon as the sugar is largely consumed the enzymns attack the protein contents of the plant cells, and these continue to be destroyed to a certain extent throughout curing and fermentation. With the decomposition of the protein there is a formation of "amido" compounds.

During the fermentation there is a loss of nitrates, a decrease in the nicotine contents, and also in the amount of tannin. A thoroughly sweated tobacco contains but a trace of tannin and thus the bitterness, due to its presence in half-sweated tobacco, is removed.

There is also a disappearance of a portion of the fat contents. A large amount of fat, or of protein, will create products during combustion that will be destructive of the finer aroma, and one of the benefits of fermentation is that it largely does away with those compounds.

There is also a decrease of the resin and gums in fermentation. In fact, one of the methods of determining how far the fermentation has proceeded is to feel the leaf and note the presence or absence of gum. The resins and gums seem to bear a close relation to the aroma. It is probable that they are split up into other products that are aromatic. It may also be that the products arising from the decomposition of nicotine have a large part in the production of aroma. The aroma of a cigar does not appear to be based on the presence of a large amount of nicotine in the cigar, in fact, a cigar rich in nicotine may be poor in aroma; the aroma does, however, seem to be in some way related to the amount of nicotine that was in the tobacco and that has disappeared in the process of fermentation. Nicotianine has been supposed to have something to do with the aroma; but this product, which may be formed from nicotine in the sweat, does not appear in all tobaccos, some of which are rich in aroma.

It has recently been shown that during the process of smoking, an ethereal oil is formed from certain products of the sweat, and that to this oil is due a portion of the flavour of tobacco smoke.

Citric, malic and oxalic acids are present in the cured leaf, but not in as large quantities as in the green leaf. The citric and malic acids may be partially transformed in the fermentation to acetic and butyric acids; this is particularly true of Perique tobacco which is cured in its own juice. These acids certainly have something to do with the aroma. The presence of the malic salts is supposed to make the leaves more soft and pliable and to give life and elasticity. This is due to the hygroscopic action of these salts. Stored under similar conditions, leaves rich in malic salts may contain three per cent. more water than leaves poor in those salts.

Cured or fermented tobacco is said to have "grain"; this grain is a product of the oxidation in the sweat, and some manufacturers consider the presence of a well-developed grain as an evidence of good tobacco. It is at least an evidence that the tobacco has been thoroughly cured. This grain is due to the formation of crystals of calcium oxalate.

During the curing and fermentation processes there often appears on the leaf an efflorescence called by tobacco men "saltpetre," or by some "light mould." This is due to the presence of potassium, sodium, magnesia, calcium and nicotine salts. These salts may be present in the leaf in excess and are forced to the surface in the process of curing or fermentation. Their presence greatly injures the saleability of the tobacco. A spray of a four per cent. solution of acetic acid, or a weak vinegar, will remove them, although more appear later. Sometimes these salts are taken off with a light brush. It is supposed by some that the presence of these salts is due to an excess of their basic elements in the soil, or to the fertilizers used. This may be true, for

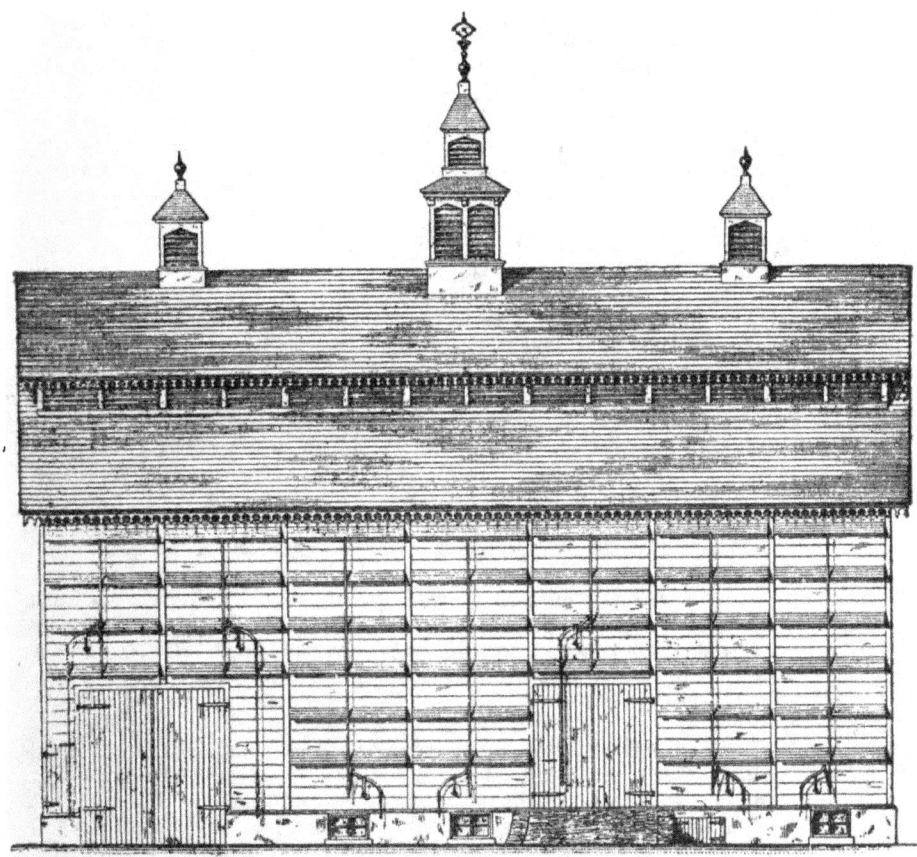

A PENNSYLVANIA BARN (COPIED FROM "10TH CENSUS," U.S.A.).

it is known that plants will take up mineral salts in excess of their requirements if those salts be abundant in the soil.

All tobaccos will not go through an equal degree of fermentation. In some the fermentation proceeds very slowly, in others rapidly. In some tobaccos the fermentation may continue a long period, while in others it may be completed in a short time. This difference is due to the presence of oxidizing enzymns in greater or less proportion. Differences in soil, climate and cultural treatment will bring about great differences in the enzymns present in the leaf and thus naturally a difference in the fermentation of each tobacco. The methods to be pursued in the fermentation of each new tobacco will have to be worked out for the special case involved.

BUILDINGS; THE CURING BARN.

The present curing barn is the result of an evolutionary process. At first the tobacco was hung in the sun upon bushes to dry; the next step was the curing rack; after this came the hanging of the tobacco in any old building that could be used; then came rough log buildings and sheds for the tobacco crop alone; and finally came the construction of buildings where the conditions could be governed by ventilation. In the use of fire there have been gradual changes from a smoky smudge to the use of charcoal, and then to the adoption of the flue system where smokeless dry heat is used. We have no reason to think that we have reached perfection in the construction of the curing barn, or in the method of applying the heat. The ideal curing barn is to be the product of the next generation.

The perfect curing barn is one that is so constructed that the tobacco may be handled with the least labour, and so ventilated and heated that the temperature and moisture conditions are absolutely under control. The heat should be applied in a form that is free from all odours, and the system used should be such that desired temperatures can be maintained, with the minimum of labour and with no unnecessary waste of material. It should be constructed out of the material most easily and cheaply obtained in the locality, providing that material has the qualities necessary to a curing barn.

In America, timber is the material most easily obtained and out of which the barns are constructed. There appears to be no reason why brick buildings cannot be used. Temperature and moisture conditions will not change as rapidly in a brick building as in a wooden building, and this certainly is a desirable feature. Buildings with thatched roof and even thatched sides can be employed for the curing of tobacco where fires are not used.

The main objection to the use of grass is that it would harbour the spores of moulds and rots, once these diseases have obtained a foothold in the curing barn. In Sumatra the curing sheds are thatched with the leaves of a palm. Galvanized iron is subject to too great and rapid changes of temperature to be a perfect material for the construction of curing barns, but perhaps this tendency to become too hot or too cold can be somewhat regulated by a covering of the iron with grass thatch.

The width of a curing barn should not be much over thirty feet, for that is about the width that can be conveniently ventilated. The height is regulated by the distance that it will pay to raise the tobacco while hanging it; from twenty-five to thirty feet is about the limit in this regard. The length of the curing barn is governed only by the amount of tobacco that the grower can place in the barn in a given length of time. It is not wise to have tobacco in the same barn in different stages of curing. Perhaps a good length is any distance up to one hundred feet. It is true that many fine barns are constructed that vary greatly from the dimensions here given. But these distances are the ones that the best tobacco men consider as most nearly correct. Tobacco barns are often several hundred feet long, and again others are sixty to eighty feet wide and fifty feet high, but

LOG BARN FOR FLUE CURING, IN PROCESS OF CONSTRUCTION.

PUTTING TOBACCO IN A FLUE CURING BARN, SOUTH CAROLINA.

such distances are likely to be more imposing than useful. It is better to build many small barns rather than have a few inordinately large.

The same style of barn may be used for the air-curing of all tobaccos, and may also be used for the curing of tobaccos by open wood or charcoal fires. In case of the flue-cured tobaccos, these barns would be too large to properly maintain the heat necessary for the fixing of the colour in the leaf, so that for this method of curing there must be a special type of barn erected.

Flue-Curing Barns for Yellow Tobacco.—As will be remembered the curing of yellow or gold leaf tobacco takes but four or five days, and during that time requires well regulated, and at times, intense heats. The barns are not large, being of a size that can be filled by a few people in a very short length of time, the idea being that all the tobacco in the barn should be in the same condition, and that the curing process should be started at once. These barns are usually from sixteen to twenty feet square, and in height, eighteen feet to the eaves. Throughout America they are constructed, as a rule, of pine logs dovetailed into each other at the corners. Sometimes, however, the frame is constructed of light timbers and heavily sheathed with boards. The main feature in the construction of these barns is to prevent the escape of heat required, or the ingress of cold air when it is not wanted. The barn has but one door and the tobacco is carried in through this. At the ground there are several small ventilators that can be opened at will and used for the entrance of cold air. In the roof, or in the gable end, there are ventilators for the purpose of allowing the escape of hot air and moisture. These ventilators are so constructed that they may be opened and closed at will from the ground. At one end of the building a small shed-roof is erected, and under this shed are constructed two brick arches or furnaces. From each of these arches, which extend about a third of their distance into the barn, runs a twelve-inch iron flue. These flues extend along the dirt floor to the opposite end of the barn, and then return at a slightly greater elevation to near the point where they started, where they pass through an opening in the barn, and carry the smoke into the open air. These barns are often burned. Care should be used in the construction, to avoid the placing of woodwork against the furnaces or the hot flues. There is no reason why these barns cannot be constructed of brick, and roofed with galvanized iron. In such a building, the iron should be covered with thatch to prevent the influence of the outside air upon the inside temperature. In America the cost of these barns, fully equipped with flues, is only from fifteen to twenty pounds. The reason for this is the extreme cheapness of material in the districts where this method of curing is used. Each barn will cure one filling a week, so that where there has been a succession of plantings, a barn will do service several times in a season.

Other Barns; a Florida Barn.—This barn is constructed for the curing of cigar leaf, but may be used for the curing of any tobacco with the exception of the yellow. It is ninety-six feet long by thirty-six feet wide; is sixteen feet from the sills to the plate, and thirteen feet from the plate to the ridge pole of the roof. The tier poles are four feet apart each way, so that there are four tiers with the plate tier and below, and two tiers above the plate. Where the priming or single

100 THE CULTURE OF TOBACCO.

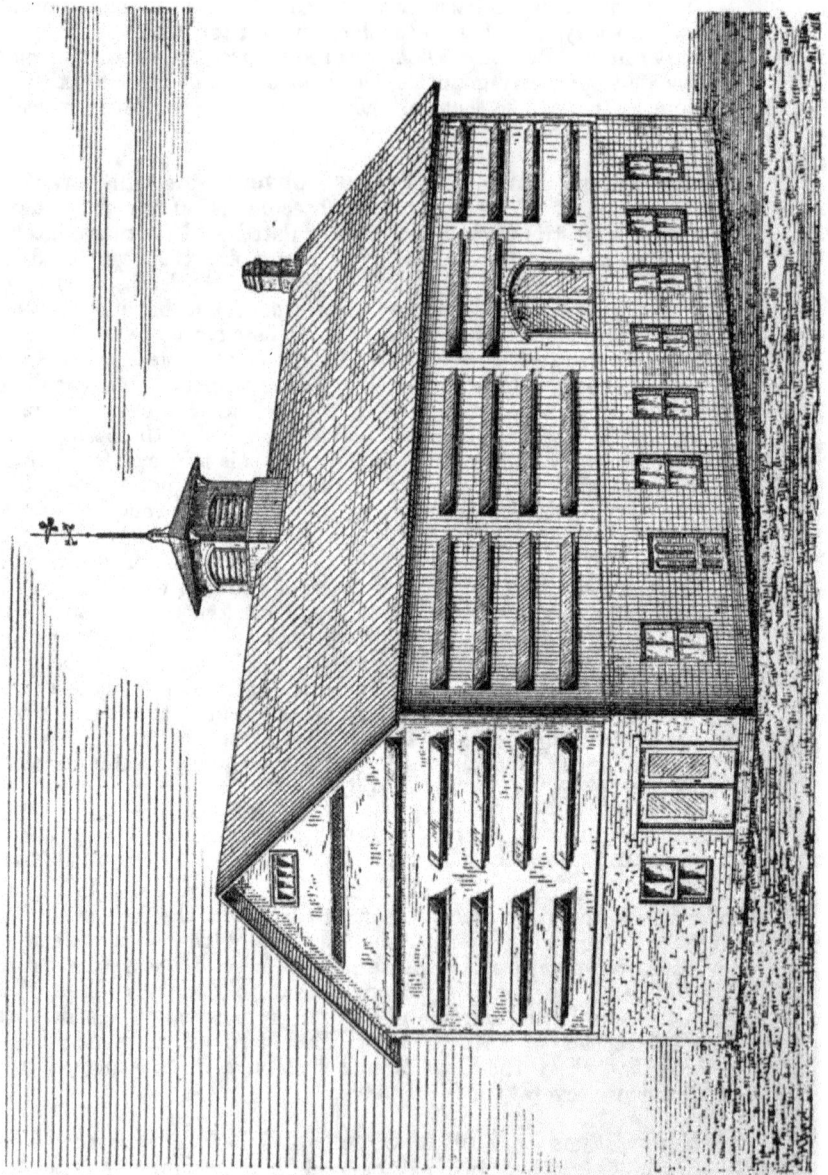

A CONNECTICUT BARN (FROM "10TH CENSUS," U.S.A.).

A KENTUCKY TOBACCO BARN; A SMALL STRIPPING HOUSE IN FRONT.

leaf system is used, the tiers are but two feet apart vertically. Where the barn has four-foot tiers, it may be changed to suit occasions by the stretching of heavy wire, so as to alternate with the original tiers. The barn has several double doors in the side, and has a series of windows or openings along both sides, that close with shutters hung from the top with hinges. These are for the purpose of ventilation. Along both sides at the base and the top there is another series of ventilators. They are long and narrow, and are closed with doors composed of one board, hinged at the upper side.

This barn is constructed entirely of timber, and rests on brick pillars. For its construction there is required twenty-one thousand feet of timber, and thirty thousand shingles, also two thousand bricks, two barrels of lime, seven kegs of nails, and a quantity of hinges, staples, and wire. The total cost in Florida when erected and before it is painted is about one hundred and twenty pounds.

Several modifications of this barn are in use. In one, the doors are at the end of the building, and the central tier poles arranged so that they may be removed, and the wagon driven into the barn. As the barn fills up, the tier poles are replaced and filled. The ventilators on both sides are vertical, and are long and narrow, being but eighteen inches wide, and the height of the side of the barn. They are hinged at the top and opened at the bottom. The vertical ventilators have one fault, and that is when open they will allow the entrance of a drifting rain. Where the nature of the construction will allow it, horizontal ventilators should be constructed. They should be but a foot wide, and may be as long as the board out of which they are made. They should extend in rows along both sides of the barn, and these rows placed one above another in every three or four feet of the side of the barn. It is also necessary to have several ventilators in the gable ends of the building, to provide for the ventilation of the peak of the barn. The ventilators should be hinged from the top, and so constructed that they may be held open at any point of elevation. They should also be adjusted so that they may be opened or closed by means of levers worked from the ground. One ninety-six foot barn will hold all the tobacco from two or even three acres of cigar leaf, or twice that acreage of other tobaccos. Where the cutting season extends over a period of two months or more, the same barn can be used for two different curings.

A Pennsylvania Barn.—This barn is a very elaborate one. It is forty-one feet wide and eighty-four feet long; twenty-nine feet from the wooden floor to the plate, and about eighteen feet from the plate to the ridge pole. The room holds seven tiers of tobacco in the body and three tiers in the peak. The building is ventilated by horizontal openings four feet apart, and so arranged as to be on a level with each tier of tobacco. These ventilators are a foot wide and are arranged in vertical series of twelve feet in width, so that each series may be controlled with one lever. Half way up the roof there is a sheltered slatted ventilator, and at the ridge there are a number of ventilated cupolas.

Under the entire barn is a basement or cellar, which is nine feet clear in height. This basement is divided into two rooms, the larger one of which is used for dampening and conditioning the tobacco, which is lowered into it through trap-doors in the floor. In this room the tobacco is also bulked. The smaller room is used for the stripping

AN OHIO TOBACCO BARN; IS OF TOO GREAT LENGTH.

FIRST CLASS TOBACCO BARN, KENTUCKY EXPERIMENT STATION.

TENNESSEE TOBACCO BARN FOR CURING WITH OPEN FIRES.

and grading of the tobacco. It has tables arranged all around it near the windows, and on these the stripping is done. The room is heated by a stove in cold weather, and has two doors, one of which opens into the larger room and the other to the outside of the building. The building is constructed of the very best material, is covered with three coats of paint, and fitted with all the little conveniences possible. The cost is about eight hundred pounds. This barn could be easily used for other purposes if the tobacco industry should at any time fail.

General Facts.—Each locality, and in fact, nearly every farmer, has a different form of curing barn, and this difference is based on the judgment, or mis-judgment, of the individual as well as on the material available and the amount of money at command. Each tobacco grower should be capable of designing his own barn, the points to be borne in mind being that the building should be so constructed that it is practically air-tight when closed, and easily ventilated when necessary. One side of the building should be facing the prevailing breezes so that they may pass through the ventilators. The exception to this is in the case of the flue curing barns. These should be so placed as to be sheltered from all winds as far as possible, for a heavy wind will greatly lower the temperature on the windward side of the barn and prevent an even curing.

Each barn, where air-curing is used, should be furnished with a stove so that the temperature of the room as well as its humidity may be regulated in cold or wet weather. The best method of applying heat would be by means of steam confined in pipes. Such a system would be expensive to install but would give greater satisfaction than any other. The heat could be evenly distributed to all portions of the building and the temperature could be regulated to within a degree. Coal could be used as a fuel in the furnaces and the labour of maintaining the fire would not be at all exacting. Electric thermometers and automatic devices could be secured for the shutting off or turning on of the steam when the temperature of the room went too far in either direction. One boiler plant could be established a short distance away from any other building, for the generation of steam for all the barns. The isolation of the boiler would reduce the risk of fire in the curing buildings. The steam could be turned into the barn whenever the weather was too dry or the air too low in humidity. More and more it is being found that it is not wise to depend altogether upon weather conditions for the curing of tobacco; nor does it produce the best tobacco to go to the other extreme and cure the leaf rapidly by means of intense heat or open fires. The method that gives the highest quality of tobacco is the one where the tobacco is permitted to slowly cure, as long as the weather is favourable, and where heat is used to regulate the conditions when the weather is unfavourable. It is not at all improbable that barns with perfect ventilation and heated by steam will soon be common in the best cigar leaf districts.

The tier poles should be strong enough to bear the weight of a man. The usual size in America is two by four inches. The curing sticks are seldom more than four feet long, which is the most convenient length for handling. These sticks are about an inch in diameter; often they are not square, a cross section being oblong in form; at other times they are triangular, and again one edge is bevelled. Small bamboo poles or canes could be used for this purpose.

A NEW YORK TOBACCO BARN (PHOTO U.S. DEPARTMENT OF AGRI.)

108 THE CULTURE OF TOBACCO.

INTERIOR OF A TOBACCO BARN.

PACKING HOUSE ON A SOUTH CAROLINA FARM.

Stripping, Grading or Packing Houses.—These may be a portion of the tobacco barn, as was the case in the Pennsylvania barn described, or they may exist as separate buildings where the tobacco from all the barns is handled and stored. The packing house should be so arranged with poles that the tobacco can again be hung up when necessary. The moisture and temperature conditions should be as much under control as those of the curing barn; for this reason, the building must be plentifully supplied with ventilators and should have a stove for heating. The best packing houses have a basement where the tobacco may be hung when the weather is dry, to bring it into order or condition for handling. That portion of the building where the grading is done must be well lighted, but not with the direct rays of the sun. The colour of tobacco cannot be correctly determined in a bright sun. For this reason the grading rooms in American packing houses have their windows on the north side. In South Africa the least direct sunlight would be on the south side of the building during the greater portion of the year.

Fermentation Houses.—These buildings are not used by small farmers. The only growers who do their own fermentation are the large companies. To make this business profitable and to pay the large salaries demanded by the experts, requires that a large quantity of tobacco be fermented and packed. The buildings are usually very large, constructed of brick, and fitted with every modern convenience. The whole building, and particularly the fermentation room, is fitted with steam heat, and the steam is so arranged that jets may be turned into any portion of the building to increase its humidity. The building has a storage room where the unfermented tobacco is stored, one or several fermentation rooms, a well-lighted grading room, a packing room, and a storage room for the packed tobacco, as well as any other additions the circumstances suggest.

PACKING OR PRIZING OF ALL TOBACCOS EXCEPT CIGAR LEAF.

After tobacco has reached the proper condition, it is ready to be packed for storage or shipment. All tobaccos intended for other purposes than the manufacture of fine cigars (tobacco for stogies and Regie cigars is handled as are ordinary tobaccos), are packed in hogsheads. Some of these may be sixty inches high and from forty-two to forty-eight inches in diameter; others are fifty-six inches high and forty-eight inches in diameter. Still others are forty-eight inches high, thirty-six inches in diameter at the smaller end, and thirty-eight inches at the larger end; this difference in width is to allow the package to be easily removed, and replaced when the tobacco is sampled. There is no bulge in the centre of a tobacco cask.

The staves of the hogshead are made of white oak or any tough hard wood, although recently the tendency has been to use cheaper woods, and pine casks are often seen. The hoops are usually of hickory or elm. In order to save freight the material is received in the "flat" by the shipper, and the erection of the hogshead is done at the same place as the packing. The price of a good cask is from five to seven shillings.

A MAN-POWER SCREW FOR PRIZING TOBACCO

A SOUTH CAROLINA WAREHOUSE FOR SALES OF LOOSE TOBACCO.

In filling a cask the butts of the "hands" are kept to the outside. Two courses are run across the bottom of the cask, with the tails of the hands interlapping, the butts being in nearly a straight line. The spaces left at the sides and ends of the courses are filled, and two more courses run at right angles with the first, and this process is continued until the cask is filled. The man in the cask has a small board to kneel upon as he works, and this he changes from side to side as necessary. At intervals during the filling the contents of the packages are compressed. The pressure applied depends upon the weight of tobacco that it is intended to place in the cask. Hydraulic pressure is at times used, but generally a screw worked by steam or man power is the method employed. In past times a lever made out of a strong piece of timber was the only thing that the packers had for the purpose. This lever could be used in emergencies to-day, although it is not as satisfactory as the screw.

The weight of tobacco placed in a hogshead varies greatly, and depends upon the class of tobacco, the amount of pressure that it will endure without injury, and upon the market catered for. A greater weight of the low grades and trash is placed in a cask than of the fine grades. From six hundred to a thousand pounds of the lighter tobaccos like the Yellow leaf; up to sixteen hundred or two thousand pounds of the very heavy dark leaf are packed in one cask. Intense pressure causes tobacco to darken. The dark tobaccos intended for the West Coast of Africa are packed so solidly, that when the cask is removed the contents appear almost solid, rather than as an aggregation of leaves. In some places, because of convenience in handling, half hogsheads are preferred. This appears to be the style of package desired by the South African market.

There is no very good reason why tobacco should not be packed in square or oblong cases, only that the cask may be rolled, and is easier to handle than the case. The cask has also been established upon the markets, and tobacco packed in cases would at first be under the disadvantage of not conforming with long-established customs.

MARKETING TOBACCO IN AMERICA.

The various sections of America have different methods of disposing of the crop. Throughout the tobacco belt of Virginia, the Carolinas, and a portion of Tennessee, the tobacco is sold loose, or without being packed.

Each town in the tobacco belt has large warehouses for the purpose of handling and selling this loose tobacco. These warehouses are lighted by means of windows in the roof, so that there is no difficulty in determining the colour, shade and condition of the tobacco on the floor. They are also so constructed that wagons may be drawn in on to the floor, and the tobacco loaded and unloaded with as little labour as possible. For bringing the tobacco to town, the farmers often have covered wagons, and always have some way of covering the tobacco with canvas or blankets, so that it will be protected in wet weather, and kept from becoming too dry in hot weather. To be in proper condition

114   THE CULTURE OF TOBACCO.

INTERIOR OF A SOUTH CAROLINA WAREHOUSE; TOBACCO IN PILES READY FOR THE SALE.

SELLING TOBACCO.

for the sale, the tobacco must contain just enough moisture so that it will not break in handling, and so that it will stretch out smooth like a kid glove when placed under tension.

As soon as the farmer brings in his tobacco it is unloaded, and each grade weighed and placed in a pile by itself. The weights are recorded in a book, and each pile is labelled with a card bearing the seller's name and the weight of the pile. At a fixed hour the tobacco is auctioned off, the auctioneer passing rapidly from pile to pile, and the buyers crying their bids. As soon as the bidding lags, or the highest bid appears to have been reached, the pile is declared sold to the highest bidder. The buyer calls to the clerk to affix his private grade mark to the label, and a record of the transaction is made in the warehouse books; in the meantime, the next pile is being sold. Very shortly employees of the buyers load the tobacco into baskets, place it on wagons, and haul it away to their packing houses, so that nearly as soon as the last pile is sold, the warehouse is cleared of tobacco, and ready to be filled again.

The sales of the different warehouses are arranged in succession, so that as soon as one sale is completed the buyers move on to the next warehouse. Often there will be two sales in the one warehouse during the day. During the busy season the sales sometimes continue from early in the morning until late at night, and as much as fifty or sixty thousand pounds may be sold in one warehouse, so that in a town like Danville, Virginia, with nine warehouses, a large quantity of tobacco may be disposed of in one day. The Danville market sells from forty to fifty million pounds of loose tobacco during a season.

The buyers, who are representatives of manufacturers, exporters, and speculators, must rapidly determine the value of a pile of tobacco, for the sale is usually only a matter of seconds. To do this they use all their senses, particularly those of sight, touch, and smell. They must mentally determine the length, width, and texture of the leaf. The texture is ascertained by feeling the leaf with the hand, and the stretching of it is to determine its elasticity. The odour must be taken note of and attention paid to the matter of moulds. The extent to which the tobacco smells of smoke must also be noticed and allowance made for it, as well as for any over-weight due to excessive moisture. The pile is also rapidly examined to see if it is uniform in grade. Rival buyers often try to run the bids of each other up very high, and then let the tobacco go to their competitor to lessen his margin on the finished product. Several buyers will often clique together and attempt to control prices. The prices of any one market cannot be kept lower than those of other markets for any length of time, for the farmers will soon change their selling point. The buyers must settle in full with the warehouseman some time during the day of the sale, or be refused further privileges on the market,

The farmer is charged a warehouse fee of 7d. a pile and a weighing fee of 7d. a hundred, as well as a commission of two and a half per cent. on all sales. When the tobacco is delivered at the warehouse the farmer may give to the warehouseman his reserve price, and the sale is not made unless the bidding reaches the figure named. The farmer himself may follow the sale and refuse to let his tobacco go unless the price is satisfactory. As soon as the sale has been made the farmer may go to the warehouseman and receive a cheque for the amount due

LOUISVILLE TOBACCO WAREHOUSE, WITH HOGSHEADS IN FRONT.

118   THE CULTURE OF TOBACCO.

OPENING HOGSHEADS FOR SAMPLING.

THE CULTURE OF TOBACCO. 119

SAMPLING HOGSHEADS, LOUISVILLE, KY.

to him. The warehouseman is responsible to the farmer, and the latter has nothing whatever to do with the buyer.

The buyers take the tobacco to their buildings, where it is "re-ordered" and packed in hogsheads for storage or shipment.

The loose-sales system is extending and is very satisfactory to producers for several reasons:—The tobacco does not have to be held by the grower until it is in a condition for packing; neither is he put to the labour and expense of packing; it also permits the sale of small quantities, and for that reason more grades of the tobacco may be made. To the buyer also the system is satisfactory, for the reason that he can at once see the quality of the goods, and because he is better prepared to handle and pack the leaf than is the grower.

The Package and Sampling System.—In some sections the tobacco is packed in hogsheads by the grower and sold by means of samples taken from these hogsheads. The inspection and sampling is done by inspectors governed by State laws. In some States these inspectors are appointed by the tobacco Board of Trade, and in other States they are appointed by a State official. The inspector is compelled to give a large bond before taking up his duties, and if the buyer finds on opening his hogshead that the tobacco does not grade up with the sample taken he may come back on the inspector for reclamation. All differences and claims presented are settled by a board of arbitration, the arbitrators being three in number and appointed by a committee of the Board of Trade. One of the arbitrators is a warehouseman, one a buyer, and the third is selected by these two. No warehouseman or inspector is allowed to buy or to participate in the profits of any sale.

When a hogshead of tobacco is placed in a warehouse for sale, it has its end removed, is then turned on its head and the cask lifted off the tobacco, which retains the shape given it by the hogshead. A sample is then made by the inspector, who prizes the tobacco apart with a lever and takes portions of tobacco from at least four places in the bulk. This sample is tied and labelled with the name of the seller, inspector, and warehouse, as well as with the weight, hogshead number, and date of inspection. The sample is then sealed, so that it may not be tampered with, and the tobacco is sold on the merits of the sample. The owner of the tobacco is given, as soon as it is inspected, a manifest with all the data concerning the package marked upon it; this manifest is negotiable. An inspection fee of one shilling and a sampling fee of three shillings are charged; this pays for the labour of opening and recoopering the hogshead. A warehouse storage fee of six shillings for the first four months and of fivepence a month thereafter is charged. If a package has been stored until the accumulated storage fees have eaten up the value of the tobacco and no settlement is made by the owner, the tobacco is sold by the warehouseman to cover his claims.

If a tobacco is found to have been packed with the intention of defrauding, the inspector is compelled to give information to the Grand Jury. False packing of tobacco is called "nesting" by tobacco men. If the cask or hogshead is found to be in bad condition it is replaced at the expense of the owner.

A large percentage of this hogshead tobacco is sold by auction upon the merits of the samples, and for this the auctioneer receives from sixpence to a shilling a package. If placed in the hands of commission men the commission is usually two and a half per cent.

THE CULTURE OF TOBACCO. 121

SELLING TOBACCO IN LOUISVILLE, KY.

Sales of Cigar Leaf.—The better grades of cigar leaf will not stand the rough handling given tobacco at loose sales, neither will they endure pressure and treatment of packing in hogsheads. The buyers who are middle men, known as leaf dealers, visit the farmers and enter into agreement with them for a portion or all of their crop. These contracts are usually verbal, and are very likely to be broken if any changes in the market take place between the time of sale and the time of delivery. A written contract and a payment of a portion of the contract price should be insisted upon.

When this tobacco is purchased by a leaf dealer it is delivered at his warehouse, where he grades, ferments, and packs it in cases for shipment. After the tobacco has become of the proper age and condition, the cases are opened and samples taken, and by these samples the leaf dealer sells his tobacco to brokers or manufacturers. The cases in which this leaf is packed hold about three hundred pounds of tobacco.

A grower could grade, ferment and pack his own tobacco if he wished, but he is usually without the skill required for conducting the fermentation, and he is not likely to have sufficient tobacco of any one grade to make it worth while for the manufacturer to directly visit him.

The large tobacco-growing companies of Florida combine the functions of growers, leaf dealers and packers. They ferment and pack all their own tobacco as well as that of the smaller growers. This tobacco is packed in bales, and is sold by a private bargain to buyers who visit the warehouse and examine the tobacco on the spot.

Leaf Dealers.

These middle men play an important part in the tobacco industry. By purchasing the farmers' tobacco they give them a ready market and relieve them of the necessity of holding their stock. Through devoting their attention to the business, the leaf dealers are able to develop new markets and handle the tobacco in a manner to suit their requirements. By handling large quantities of leaf, they are also in a better position to secure favourable consideration from manufacturers and foreign buyers, than would farmers themselves. It is doubtful if a thrifty tobacco industry could be developed in any locality without the aid of the middleman. Where the tobacco industry has become thoroughly established and the manufacturing branch has been consolidated (as has been done by the Tobacco Trust in America) the independent middleman is superfluous. But in all newly developing localities the presence of the middleman, even though he make enormous returns for his labours, is to be encouraged.

There is apparently no reason why an association of planters could not be incorporated for the purpose of conducting the business usually carried on by leaf dealers, tobacco brokers and packers. Such an association would be the way out of the difficulty in localities where the business is not being conducted by private enterprise.

Re-ordering and Stemming Tobacco.

Tobacco, as it comes from the curing barn or the warehouse floor, is not ordinarily in condition for shipment in hogsheads. The body

A STEMMING AND RE-ORDERING ESTABLISHMENT IN S. CAROLINA.

124 THE CULTURE OF TOBACCO.

INTERIOR OF A RE-ORDERING ESTABLISHMENT.

TAKING TOBACCO AWAY FROM A RE-ORDERING MACHINE; FACTORY OF PATTON AND CO., DARLINGTON, S.C.

126   *THE CULTURE OF TOBACCO.*

A LARGE RE-ORDERING MACHINE (PHILA. TEXTILE CO.)

A SMALL RE-ORDERING MACHINE (PHILA. TEXTILE CO.).

of the leaf may be so dry that it will easily crumble, and at the same time the stem, while apparently dry, will contain enough moisture to furnish food for the growth of mould.

In every tobacco market there are certain men who make a business of "re-ordering" tobacco and packing it for shipment. The manager of a re-ordering plant may also be a leaf dealer, or he may only handle tobacco for leaf dealers. Several methods of bringing the tobacco leaf into the proper condition are practised. The first and oldest method is where the tobacco is hung on racks in large rooms and brought into condition by protracted hanging and a proper regulation of the ventilation. The tobacco is first allowed to become so dry that the stem will snap between the fingers, then the windows are opened on some moist day, and the leaf permitted to absorb just enough moisture to keep it from breaking when handled, or until the fibre of the leaf becomes pliable while the stem is still dry enough to break. It is then in condition to pack. If it is desired to pack the tobacco when the weather is very dry, it may be conditioned by being hung for a few minutes in a box filled with hot steam. This system of "air ordering" the tobacco calls for the use of a large building, as well as the tying up for a long time of the capital invested in the tobacco. This, in addition to the insurance, will in the end make this method more expensive and less satisfactory than any other method.

In the second method the racks are built on trucks, and these trucks run into a steam-heated room and the tobacco kept there at a temperature of from 150° to 160° until it will crumble in the hand. It is then conditioned by jets of steam turned into the room, and is packed in hogsheads while still hot. These rooms are fitted with steam fans, so that the hot air and the steam are evenly distributed. This method is perhaps the best one to adopt where quantities less than two or three hundred thousand pounds are racked annually.

The third method, which is called "machine ordering," is the one generally used where large quantities of tobacco are handled. At one end of the machine the tobacco is placed on an endless wire belt that carries it through the different compartments and finally delivers it to the packers.

The tobacco is first carried through two very hot steam-heated chambers to place it in a thoroughly dry condition. The first chamber is not heated as hot as the second, for the reason that it might cook the tobacco which contains some moisture. The second chamber is heated to 160°. From the second chamber the tobacco passes into a third where it is partially cooled by a current of air, and then into a fourth where it is by the use of steam made pliable enough to handle. As soon as the tobacco comes out of the end of the machine it is packed in hogsheads while still soft. When the tobacco is packed it must be moist enough to handle without breaking, and yet contain so little moisture that it will appear perfectly dry when cold. A machine has a capacity of from fifteen to twenty thousand pounds of tobacco a day and has this merit, that within an hour after a tobacco has been brought into the building it may be packed and shipped. The use of a machine makes it possible to do the work in a building but very little larger than the machine. An equipment of this kind, exclusive of the building and engine, costs about four hundred pounds in America.

STEMMING TOBACCO FOR EXPORT TO ENGLAND.

Tobacco intended for the English market is packed as dry as possible, because of the high duty in that country. In the case of most tobaccos for this market the weight is reduced by the removal of the stems.

Stemming or Stripping.

The tobacco is placed in a very moist condition, and the stem, or all of it with the exception of a few inches at the tail end, is removed by one motion of the hand. The leaf, or strip as it is then called, is then run through the ordering machine and is packed in hogsheads in as dry a condition as possible. Most of this stemming is done by women and children, who are paid by the quantity of work completed. An expert stemmer can remove the stems from two hundred and fifty to three hundred pounds of tobacco in a day. The stems average from twenty to twenty-five per cent. of the total weight of the leaf. There is also a wastage resulting from broken leaves of about five per cent. of the total weight. This broken tobacco, or scrap as it is called, is sold to tobacco factories for the purpose of manufacturing granulated smoking tobacco. The stems are used for making sheep dip and fertilizers. Stems are also crushed and cut up for mixing with the cheaper grades of pipe tobaccos, and they may also be used for the purpose of making cheap snuff.

Machines have been invented for the stemming or stripping of tobacco. None of these machines have as yet reached perfection or come into general use. It is very probable that a machine that will do the work in a satisfactory manner will be the product of the next few years.

SOME OF THE CHEMISTRY OF THE TOBACCO PLANT AND ITS RELATION TO THE FERTILIZERS USED.

Tobacco is fertilized for the production of quantity and for the development of quality. Fertilization of the soil for the production of quantity is a simple matter, and no more difficult than the fertilization of the land for large crops of wheat, hay, or maize. But the quality of the tobacco produced, lying deep in the shadow of nature's secrets, and being of vastly more importance than the quantity, is a far more difficult thing to control. Tobacco that does not meet with approval for human consumption is unfortunately of little value for other purposes. The quality of the potato can be largely decided by a determination of its starch content; the quality of maize may be estimated by chemical tests to determine its starch, protein, and fat percentage; but the quality of tobacco cannot be determined by a chemical analysis, but only by the senses of man. Colour, texture, size, aroma, flavour and combustibility are the points by which the quality of tobacco is estimated.

The different salts and compounds in the leaf, both organic and inorganic, are known to have a great effect on all of these different qualities, but the action of each one of them is as yet not completely understood. Some facts regarding the action of the different elements have, however, been obtained, and until these are understood by the tobacco planter, he will not be able to exercise his best judgment in the matter of the fertilization of the soil. The organic compounds seem to bear the closest relation to the aroma of the tobacco, while on the inorganic salts depends largely its

combustibility. It has been determined by experiments in Europe, America, and Japan—

1. That chlorine is injurious to the burning qualities of the leaf, and that the excess of sulphuric acid is also injurious in the same way.

2. That the presence of a large amount of potash in the tobacco greatly improves the burn, but that the combustibility is not proportional to the percentage of potash present, but is dependent on the amount of potash in excess of the amount required to combine with the mineral acids such as chloric and sulphuric acids, and that the potash, when in the form of a carbonate, gives the best results.

3. That the alkalinity of the ash has a marked effect for good on the combustibility.

4. That a certain amount of lime, where potash is deficient, may combine with the free acids, and improve combustibility, but that an excess of lime may be harmful.

5. That the small percentage of iron and aluminium oxides that may be present in the leaf have no apparent effect on the combustibility.

Tobacco plants require certain quantities of several different elements, as lime, potash, sodium, iron, sulphur, phosphorus, chlorine, nitrogen, &c., to complete their development, but it will be seen that the presence of an excess of some of the elements, as sulphuric and chloric acids, is harmful to the commercial product, and that the presence of other elements, as potash, in excess of the real needs of the plant is beneficial to the cured article. Therefore, the conclusion must be that in the use of fertilizers for the stocking of the soil with those elements it usually lacks (as potash, phosphoric acid, and nitrogen), materials should not be used that have as impurities elements likely to be injurious, such as chlorine, which is ordinarily present in the form of common salt (sodium chloride).

A large percentage of nitrogen in the soil increases the production of albuminoids, and a large percentage of albuminoids in the leaf is considered objectionable, for they cause the leaf to burn badly, and to have a disagreeable odour. An excess of nitrogen also tends to the production of a thick, coarse leaf, and such a leaf will burn less readily than a leaf of finer texture. The heavier and coarser the leaf the greater is the tendency to the formation of nicotine. The accumulation of nicotine and of an excess of albuminoids is more injurious to the quality of ordinary smoking tobaccos than it is to cigar leaf, for those compounds are largely reduced and transformed during the process of fermentation that the latter leaf goes through.

Phosphoric acid is usually deficient in tobacco soils, and must be supplied, but an excess of phosphoric acid in the leaf has a tendency to hasten the ripening process, and may injure the burning qualities of the leaf.

No fixed formulæ can be given for the fertilization of tobacco, for the reason that the elements, and the amounts required, will differ with each change of soil and climate. The grower himself must determine, from his knowledge of the soil and its requirements; of the tobacco plant and its nature; and of the action of different fertilizers and their relative cost, as to the extent and composition of the manuring that he will give his land.

Stable manure is one of the best fertilizers where a large, coarse leaf is desired, but where the finer types are to be produced, it must be used in lesser quantities and be supplemented with commercial fertilizers. Stable manure is nitrogenous, and if used in excess will make the leaf produced thick and strong. However, stable manure has a value beyond that of its apparent fertilizing value. By its decay in the soil it warms it and stimulates plant growth. The decaying of manure also assists in the dissolution of different fertilizing elements and in placing them in a condition to be of value to the plant. It also encourages the action of the beneficial nitrifying bacteria and improves both the drainage and the water-holding capacity of the soil by the addition of humus. Humus is decayed animal or vegetable matter. The greater the amount of humus in the soil, the greater the tendency to the thickening of the texture and the darkening of the colour of the leaf, so that there is a point in the production of fine tobacco where the addition of further humus is not desirable. Swamp or vlei lands are, as a rule, rich in humus, and the humus may, or may not, be rich in nitrogen, that being a matter dependent on the nature of the vegetation that formed the humus, and upon the drainage conditions that the humus has been subjected to:

Humus and nitrogen may be added to the soil by the ploughing under, or the decay upon it, of nitrogen gathering crops, as peas, beans, clover, beggar weed, and other leguminous plants. The action of these upon the soil is somewhat similar to the action of stable manure. Leguminous crops have this further virtue, that they draw their nitrogen largely from the atmosphere, and that their roots attack stores of plant food beyond the reach of the tobacco plant, and bring these materials near the surface of the soil, where they are left in an easily digestible condition for the roots of the tobacco plant. But as these crops have the same action in thickening and strengthening the tobacco leaf as has stable manure, they should be used sparingly for the finer types of leaf. The growers of Bright tobacco, the chief merit of which lies in its colour, find that the tobacco grown on the land immediately after the ploughing under of a leguminous crop is deficient in texture and colour. However, the soil must be kept stocked with a certain amount of nitrogen and humus, and the method of these growers is to plant some other crop like cotton or maize upon the land the first season after the addition of the leguminous crop, and to partially exhaust the nitrogen by means of it. A humus that is not so rich in nitrogen may be added by ploughing under a crop of weeds, or rye, or buckwheat. The Bright tobacco planters find that if they allow their land to grow up to grass and weeds for a year, and plough this under, that they have the land in the best condition for a fine crop of tobacco. The allowance of this length of time would hardly appear necessary, for the same results might be obtained by the sowing of a rapid growing, non-leguminous, catch crop like buckwheat and the loss of the land for a year be avoided.

Nitrogen may be added to the soil in other forms, as nitrate of soda, sulphate of ammonia, or nitrate of potash. This latter salt has the additional advantage of supplying the potash as well as the nitrogen for the crop. Where cotton is produced the use of cotton seed meal (the product left after the extraction of the oil) is to be highly recommended. Any decaying vegetable or animal matter contains a certain amount of nitrogen. The free use of animal products, as dried blood, or "tankage," is not to be advised, for these fertilizers may contain large amounts of chlorine, to which element, we have before seen, the tobacco plant is extremely sensitive. The use of available stable manure and cotton seed meal should furnish the basis for the nitrogen supply and the deficiency be made up by the use of nitrogen salts.

Potash may be added to the land in the form of ashes from wood fires. These ashes will have their value largely impaired if exposed to rains and allowed to leach before being placed on the field. In cotton oil mills the hull of the seed is used as fuel for the boilers, and the ash left is of considerable value as a fertilizer for the supplying of potash and phosphoric acid to the soil. The potash fertilizers employed usually come from the mines of Germany. They consist of a number of different salts, several of which, as "kainite," contain large amounts of chlorine in the form of common salt which would be injurious to the quality of the tobacco. High grade sulphate, sold under a guarantee to contain no chlorine at all, is one of the best forms to be used for the tobacco crop. The very best fertilizer is the carbonate of potash, for this salt contains no elements injurious to the crop. It will neutralize the evil effects of the injurious elements already in the soil, and its presence will be greatly beneficial to the burning qualities of the tobacco.

Ground bone and ground phosphate rock are two of the forms in which phosphoric acid may be applied. When these have been treated with sulphuric acid they are known as dissolved bone and superphosphate. The use of the acid makes the phosphate more

OLD "DARKIE" APPLYING FERTILIZER TO COTTON; SAME METHOD OFTEN USED WITH TOBACCO.

soluble and more available for the plant, but, at the same time, the percentage of actual phosphoric acid to the ton of fertilizer is reduced. The sulphuric acid in the superphosphate may prove injurious to the combustibility of the leaf, and surely will if there be any great excess of it present. The better form, after all, may be to apply the rock or bone in as finely ground form as possible, and let it become available through the action of the soil, the weather, and the plant juices.

Lime is an essential to the growth of the plant, but most soils contain sufficient for the actual needs of the tobacco plant. Lime's chief value to the crop lies in a secondary action. Where the soil is acid it corrects the acidity, and promotes nitrification by encouraging the action of the nitrifying bacteria. In the heavy close clay soils it promotes floculation, or the combination of the small particles of the clay into larger bodies, thus making the soil more friable, easy of cultivation, more rapidly drained, aerated, and better adapted to the production of roots and the development of plant life. It has also the power of making certain food materials in the soil, that have become insoluble and non-available, of use to the plant by displacing them from their present combinations. This is particularly true of lime's action on potash, and not unfrequently lime is given the credit of being of great value to the plant in the matter of plant food when its action has really been simply to make other materials available.

A crop of a thousand pounds of leaf, if the stalks be returned to the field, may be said to remove from the soil about forty pounds of nitrogen, five pounds of phosphoric acid and fifty pounds of potash. This does not mean, however, that only this amount of these elements must be available, or be added to the soil each year, to produce a good tobacco crop. The tobacco plant makes its growth in about sixty days, and fertilizers to be available to the growing crop must be in an easily soluble form. Where the fertilizers are added in forms that are not easily soluble larger amounts must be used. The tobacco roots do not reach and secure all the materials present in the soil and available; therefore allowance must be made in the fertilization for this, and also for the fact that some of the more soluble forms will be washed out by the rains and lost. The soil already contains amounts of the elements needed, and, if it be well balanced, it may not be necessary to fertilize heavily for several years. However, if the process of depletion without restoration be continued, the soil will become exhausted and of little value. It is a much easier process to maintain the fertility of the soil than it is to restore it.

Several methods of applying the fertilizers to the soil are in use. It may be broadcasted or drilled into the field so as to be evenly distributed, or it may be placed in furrows made along the rows where the plants are to be set, or again, it may be placed around the plant. No one method can be recommended for all occasions. Barnyard manure is applied evenly over the field and ploughed in. This should be done some time before the planting season, if possible, so as to allow the decay to get well started, and to permit the heat of the first fermentation to subside. Where commercial fertilizers have to be purchased, the best results for the money may be obtained by placing them where the plant can utilize them as far as possible during the growing season. This may be done by distributing them in the furrow when the ridge is being thrown up in the ridge system of planting, or by distributing them along the marked row with a fertilizer drill, or by hand, where the level system of planting is used. Quick acting fertilizers, as nitrate of soda, or nitrate of potash, are often scattered around the plant while it is growing. This is done where the crop has received some check from which it must recover, or where excessive rainfall has washed the nitrogenous salts from the soil.

Nitrogenous salts should not be applied in any great excess of the present needs of the plant, for they will largely wash out of the soil before the next growing season and be lost. Potash is not so available, and perhaps should be applied in excess of the apparent requirements of the plant, but not in any great excess, for, while it is not so likely to leach out of the soil as is nitrogen, it will form insoluble compounds and be unavailable for the following crops. Where it has become insoluble it may be released by the use of lime. Phosphate fertilizers when in only partially soluble forms, as the ground bone and phosphate rock, should be applied greatly in excess of the present needs of the plant, for they will become slowly soluble during a period of years and be available to several crops. Where the dissolved bone and the superphosphate rock are used, a large excess must be avoided. Lime is better if added in moderate quantities each year, than if all placed on at one time, for lime leaches easily and is carried down to lower levels by the drainage water. Lime also gets in a form where it is of less use as a neutralizer of acids and a dissolver of potash combinations.

A rotation of crops is of great advantage in keeping up the fertility and condition of the land, but, in the establishment of the rotation, great judgment must be used where tobacco forms one of the links. Fertilizers containing impurities, as chlorine salts, may not be harmful to the other crops and may be even of advantage from the standpoint of cheapness, but these impurities will remain in the soil and injure the quality of the tobacco grown on the soil. The fertilization of the other crops must always be done with the effect on the tobacco crop borne in mind. Leguminous crops should have a place in the rotation and should be followed, if the conditions will allow it, with maize (mealies). A proper cultivation of the maize crop will place the soil in fine condition for tobacco. The heavy applications of manure should be used on the maize crop instead of on the leguminous crop. A leguminous crop's chief merit lies in its ability to remove nitrogen from the air and fix it in the soil, which function will not be thoroughly exercised if the plant be well fed with prepared nitrogen, as is the case where much manure is applied. The bone and rock phosphates may also be applied to the corn crop.

The value of a fertilizer must not be estimated by the number of tons applied, but by the percentage of actual fertilizing materials in each ton. Fertilizers containing large percentages of fertilizing elements, command a much higher price than fertilizers containing lower percentages, but the higher priced fertilizers are in reality much less expensive to buy than the lower priced ones. There is much less number of pounds of freight to be paid for in comparison with the actual fertilizing material received, and much less labour required in their application. Then again, where the percentage of fertilizing elements is low, something must be used as a filling and to give weight, and the material used may be injurious to the quality of the crop. Rags, woollen waste, animal matter and certain animal manures tend

THE CULTURE OF TOBACCO.   137

SLATED SHADE IN FLORIDA, SHOWING SYSTEM OF MANURING SOIL BY FEEDING CATTLE ON THE GROUND IN WINTER.

to the production of disagreeable odours and flavours in the tobacco when they are used as fertilizers.

In the production of tobacco there may be localities where it will pay better to give attention to the production of quantity rather than quality, and in such places large quantities of stable manure and nitrogenous manures may be used and no attention given to impurities.

WATER IN ITS RELATION TO THE TOBACCO CROPS.

Upon the percentage of water in the soils depend largely the colour and texture of the leaf. The value of many tobacco soils is not based so largely on their fertility as it is on the soil's ability to hold greater or less proportions of moisture, to drain off rapidly the surplus water in times of plenty, and, by capillary action, to draw up water from below in times of drought. The Bright tobacco lands and the cigar leaf lands of Florida have but little fertility, and are only used because of their moisture content and the warmth that is due to that moisture content. It has been determined that soils holding about ten per cent. of moisture are adapted to the production of wrapper leaves, while soils with a higher water content, as 20 per cent., are only adapted to the production of heavy tobacco and fillers.

SHADED SUMATRA TOBACCO, SHOWING SYSTEM OF FLUME IRRIGATION.

Water has many functions and offices in the development of a plant. Water loosens the soil and allows the delicate rootlets to continue their search for food; it permits the beneficial nitrifying bacteria of the soil to exercise their functions; aided by the plant juices it dissolves and makes available the plant food in the soil; it transports the plant food from cell to cell and from one portion of the plant to another; by evaporation it cools the plant and prevents the death of the delicate protoplasm through the action of the blistering sun; and more than this, water acted upon by the energy of the sun, combines with the carbon of the air and becomes the chief food of the plant, forming the starch, the sugar, the cellulose, and also largely the gums, oils, and acids that together make up at least 80 per cent. of the weight of the fire-dried plant. Water, by its sufficient presence in the curing plant, permits the development and action of the oxidizing enzymns, and of the different chemical changes that take place.

Water in excess prevents the action of the nitrifying bacteria, and favours the action of the de-nitrifying bacteria, smothers the roots by preventing the movement of air in the soil, and, by evaporation from the surface of the soil, cools the earth to the point where rapid plant growth cannot take place. Heavy rains also wash out of the soil a portion of the soluble food materials.

Heavy rains during the ripening of the tobacco plant wash out the desirable gums and oils and render the leaf thin, papery, and devoid of fine aroma. Rainy cold weather tends to the production of acids in the leaf, and these acids are detrimental to the action of the oxidizing enzymns to which the development of aroma is largely due. An undue degree of humidity is also favourable to the development of plant diseases and to the growth of moulds and rots in the curing and cured tobacco. On the other hand, if there be a great scarcity of rainfall, even if the plant be kept growing by the moisture in the soil, there will be an absence of the materials that produce a favourable fermentation. Heavy dews on the plant during the ripening period are greatly desired, for they aid in the production of gums, resins, oils, and other products that are necessary to the production of a good, sweet and finely-aroma'd tobacco. To produce a fine textured leaf means that the plant must be kept growing from start to finish, and to do this the plant must be supplied with water in sufficient quantities.

In no locality is the question of the water supply so well understood as in the cigar leaf belt around Quincy, Florida. Here the aim is at the production of fine cigar wrappers and any checking of the rapid growth of the plant would injure the product. The soil is sandy and can endure a rain every day for a week without suffering, and a moderate rain every three days would suit the needs of the soil exactly. But as nature does not always regulate the rainfall to suit the desires and requirements of man, and as a prolonged drought would injure a crop worth £200 an acre, expensive irrigation plants have been installed and nearly every acre of this type of tobacco can be irrigated when necessary.

The nature of the country is not such that open ditches can be used for carrying the water to the field. The greater portion of the water is confined behind expensive dams and is pumped to the fields as required.

Two methods of distributing the water are in use. By the first method the water is carried throughout the tobacco field in open

A FIELD SHADED WITH CANVAS IN CONNECTICUT.

flumes or troughs and from these distributed to the rows of tobacco. This is not an expensive system to install in a locality where timber is cheap, but the flumes interfere with the working of the field. A large number of persons are also required for the controlling of the water in the irrigation furrows.

In the second and more modern system the field is covered with a network of iron pipes at the height of eight or nine feet. Every thirty feet each way these pipes have a spray nozzle attached, so that the field may be given a thorough watering by the starting of the pumps and the turning of the water into the system. This system is an expensive one to install, the cost being about £80 an acre, without considering the value of the pumping plant; nevertheless, it is a very economical one to operate, for one man, besides the stoker at the power house, can control the water on hundreds of acres and at the same time have it more evenly distributed than by the furrow system. The piping is also out of the way, and does not interfere with the cultivation. These fields all have a framework for the support of a canvas covering, so that no new framework is required for the support of the pipes.

The irrigation of tobacco is certainly worthy of wide extension. To grow and ripen a crop with but little rainfall and to supply the moisture by artificial means is without a doubt an ideal condition. In Rhodesia there would seem to be time enough to grow and ripen a crop before the advent of the heavy rains, if water irrigation were practised. Another crop could probably be grown during the latter end of the rainy season and ripened after its cessation. Occasionally when the season ended early there might be necessity for the use of irrigation, although tobacco when ripening can get along for three weeks or a month very nicely without rain, providing the soil is sufficiently moist when the rains cease. A tobacco that has been ripened during heavy rains will be washed and deficient in aroma and may even be possessed of a disagreeable odour.

PRODUCTION OF TOBACCO UNDER SHADE.

Several years ago it was discovered in Florida that Sumatra tobacco grown in the shade of trees was of finer texture and better adapted to be used for cigar wrappers than tobacco grown in the open field. From this observation originated the idea of building artificial shade. The first experiment proved a success, and the building of shade has gone on until all the Sumatra tobacco grown in America is shade grown. From Florida the idea spread to Connecticut, and even to Cuba and Porto Rico.

The first shade consisted of a framework of timber on the top of wooden posts. The timbers were connected by heavy wires and these wires were covered with a lattice work of light wooden slats or laths. This lattice covering proved fairly efficient, but is now giving way to the use of cloth.

To erect a covering, wooden posts twelve feet long are first set in rows a rod apart each way, the posts being placed in the soil to the depth of three feet, thus allowing them to stand above the ground to the height of nine feet. These posts are then connected together at the top in one direction by means of light stringers. Heavy cable wires connect the posts in the other direction. Parallel with these

THE CULTURE OF TOBACCO. 143

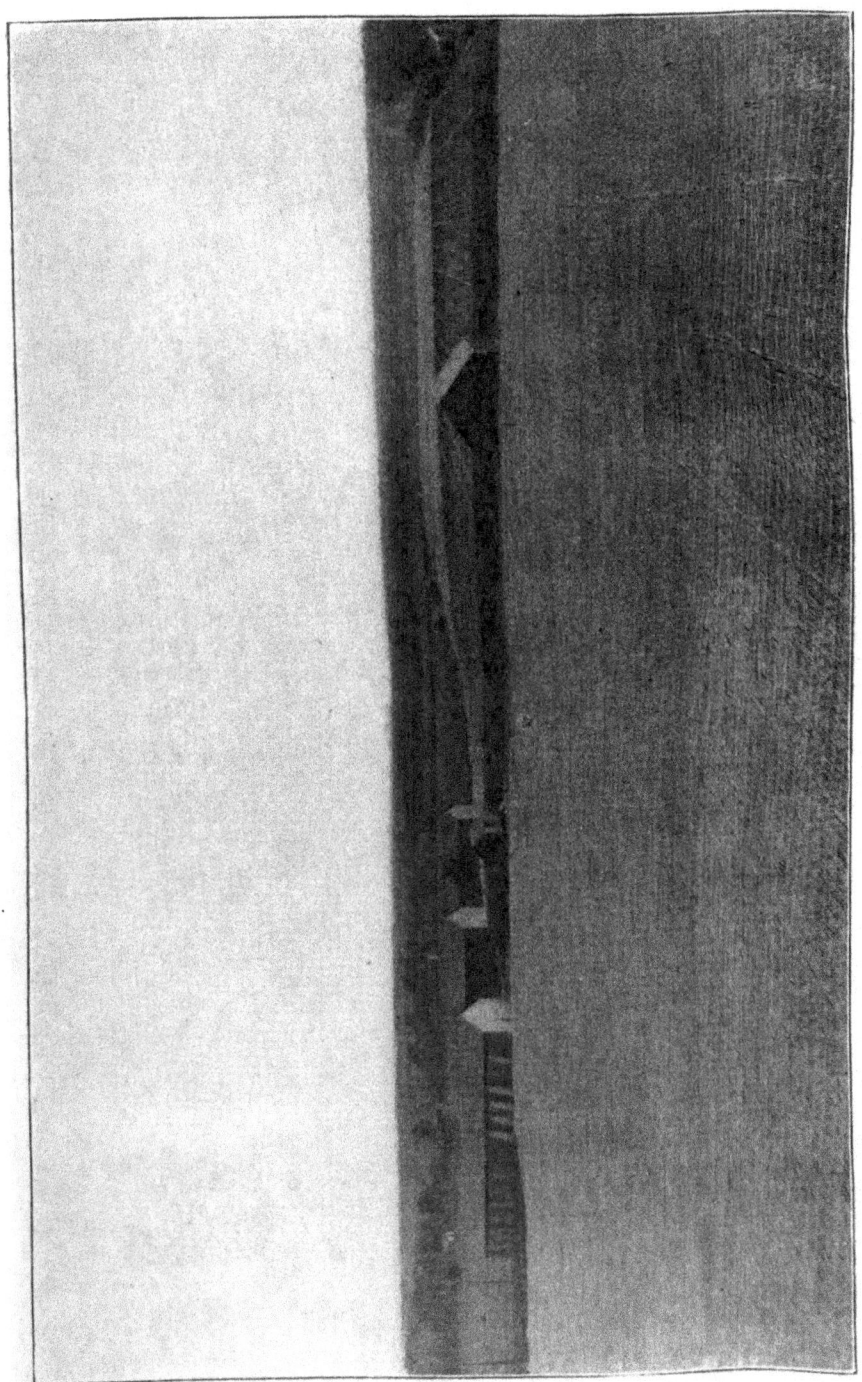

UPPER SIDE OF A SHADED FIELD, SHOWING SOME TOBACCO BARNS IN THE DISTANCE.

144 THE CULTURE OF TOBACCO.

UNDER SHADE IN FLORIDA; SUMATRA TOBACCO NICELY STARTED.

THE CULTURE OF TOBACCO. 145

SUMATRA TOBACCO IN FLORIDA; ABOUT A THIRD GROWN, AND TIED UP TO THE SHADE TO KEEP IT FROM BEING WHIPPED BY THE WINDS.

cable wires, and between them, are run wires of lighter weight. These wires are placed at a distance of five and a half feet from each other and from the cable wires, so that there are two of these light wires between each two cable wires. These wires are stapled wherever possible and the ends are securely fastened to stakes set from six to ten feet out from the edge of the outer posts. The whole field is then covered with an open mesh cloth that is much heavier than cheese cloth. This cloth has a closely woven selvage and also has strongly woven strips running through it every few feet to give it strength. The width of the cloth is five and a half yards, or just the width between the posts. The cloth is tacked on to the stringers and is supported between them by the wire. It is run the length of the field and down to the ground at both ends, where it is attached to base boards. The remaining two sides of the field are then closed in. A large gate is left for the passage of wagons, and roads are left here and there throughout the field. The framework will last from five to ten years, but it is safer to renew the canvas each season. The old canvas may be used for the covering of plant beds. In Florida the cost of the framework is about £40 an acre, and of the renewal of the canvas each year £20. In countries where cigar wrappers bring a large price and where timber is expensive and not durable, the framework for the shade could be constructed of light iron. Iron piping would do nicely for the posts.

The covering changes the climate of the field and makes the inside air more humid, as well as about fifteen degrees warmer than the outside air. The covering retains the heat during the night and permits of more rapid chemical changes in the plant. It also protects the plants from winds, frosts and hail, as well as prevents the delicate leaves from being scorched by the sun, and also very largely prevents the ravages of insects by closing them out of the field. If the insects are once allowed to enter, through carelessness in leaving openings, they will breed and multiply under the shade as rapidly as elsewhere.

With a tobacco that is valuable enough to grow by this method great care is taken in the other steps of the culture. The fertilization is reduced to a science, the soil is kept in perfect tilth, weeds and grass are not seen, and every field in some localities is irrigated. If any caterpillars are known to be in the field the plants are inspected every day and the pests destroyed. The tobacco grows very tall and would easily break down, so that it is tied to the top of the framework with twine when half grown. The plants are allowed to mature a large number of leaves, for the larger the number of leaves the thinner and finer the texture of the individual leaf. From sixteen to forty leaves are left to a plant, and where the plant is very thrifty it is not topped at all, but allowed to go to seed so as to prevent the leaves from becoming too heavy.

It is difficult to tell when this shade-grown tobacco is ready for the harvest, because it does not colour up and give the same signs of ripeness as tobacco grown in the open. A light brownish colour around the edges and the tips, and the appearance of small spots, indicate when the leaf is ready for harvesting. The Sumatra tobacco is more elastic and better for wrappers if not left to become too ripe. All the leaves on the plant do not ripen at one time, so they are pulled off when ready, and the plant is not cut. Four or five leaves are usually ready at once, and the time of harvesting a field may extend over several weeks. The leaves are taken to the barn in baskets and cured

GROWING SUMATRA TOBACCO UNDER SHADE IN FLORIDA; HALF GROWN.

SUMATRA TOBACCO UNDER SHADE IN FLORIDA; NEARLY FULL HEIGHT.

SUMATRA TOBACCO WITH HALF THE LEAVES HARVESTED (U.S. DEPARTMENT OF AG.).

by the air-curing process, in the same manner as other cigar leaf. When all the leaves are harvested, the stalks are cut down and ploughed under or hauled to a pile to decay.

About a thousand pounds of this shaded tobacco are produced to an acre, and it sells for from two shillings to sixteen shillings a pound. The value of an acre crop may be estimated at from one to two hundred pounds. In Florida this shaded tobacco is largely grown and handled by companies that produce several hundred acres a year. They are highly capitalized and can afford to pay their managers and experts high salaries. The profits are immense. The high price of this leaf is partially owing to its real value, and partially to the fact that the United States does not produce all of this type of leaf required, and the domestic article is protected by a high duty.

It is an interesting sight to walk through a large field of shaded, irrigated Sumatra, and see the magnificent plants eight or nine feet high, the large delicate leaves interlacing in every direction.

TOBACCO AS AN INSECTICIDE.

Stems, stalks, and worthless tobacco can be made to serve a useful purpose in the destruction of insect pests. The stems, thickly scattered around on the earth in the greenhouse or the plant bed, will largely prevent the attacks of plant lice. A decoction of tobacco sprayed on plants, will destroy the majority of tender skinned insects. To secure a decoction of the proper strength, boil the stems or leaf in water for a length of time, and then dilute it with enough cold water to make two gallons of the preparation for every pound of tobacco used. This can be used in stronger or weaker forms as the conditions seem to require. The fumes of tobacco if confined in a room for a sufficient length of time will largely check the ravages of many insects.

Tobacco makes an excellent dip to use for all animals for the destruction of ticks, lice and other parasites. A pound of tobacco stems to each two gallons of the dip is the proper strength.

The woolly aphis, or American blight, does great damage to the apple tree, and, while it is easy to destroy all the aphides on the body of the tree, the difficult matter has been to find something that will destroy them on the roots, where they will survive and breed new multitudes for the destruction of the tree. Tobacco stems buried in the soil around the base of the tree are the best remedy to be used in this case. If the season be dry, the tobacco applied in the form of a decoction will give the best results. Tobacco stems around the base of trees do much to prevent the attacks of the different climbing insects, and at the same time are useful as a fertilizer.

PERIQUE TOBACCO CULTURE.

Produced by Longfellow's historic Arcadians on a soil created by the muddy Mississippi, cured by unique and laborious methods, the fragrant aromatic Perique is shipped to all portions of the globe, to charm the senses of lovers of the "weed." Grand Points, in St. James's Parish, Louisiana, is the centre of the industry, and it was here that a hundred years ago, Pierre Chenet, from whom

PERIQUE TOBACCO, HALF GROWN.

MAKING PERIQUE CARROTTES.

the product derives its name, first started its culture. This locality, described in George W. Cable's "Bona Ventura," lies but a few feet above the swamps of the east bank of the Mississippi river, and were it not for the levees along that stream, the whole country would be under water for a portion of the year.

Considering its reputation and its widespread popularity, the total amount of this tobacco produced is very small. In fact, during the past two years, because of the unsatisfactory price offered, the production has dwindled to nearly nothing, but through the securing of several contracts, the industry has received new life, and this year the finished product will amount to about eighty thousand pounds, with the prospect of a large increase during the next few years.

The soil is a dark gray, well drained, fertile, friable one, containing a large proportion of very fine sand.

But one variety is grown and this, as well as the finished tobacco, is known as Perique. The plant is a rapid grower, producing leaves from medium to large in size, with a large stem or midrib. The leaf is tough, gummy and elastic, and when first dried is dark brown in colour.

The seed beds are made either in an open space in the forest, or on the sheltered side of a building. Several months before sowing time the bed is thoroughly fertilized with stable manure, and well worked up; after this several re-workings are made before sowing season. The bed is never burned.

When the land has been well prepared, the plants are set out on ridges about three feet apart, and are placed two to three feet apart in the rows. The cultivation, worming, and suckering are about the same as for other pipe tobaccos. An average of from twelve to fifteen leaves are left on the plant.

When the plant has assumed the speckled appearance indicative of ripeness, it is harvested by being cut off near the ground with a hatchet, the sand lugs being left on the field. Heavy dews are greatly desired during the ripening period, the moisture of the leaf causing the production of a large amount of gum and rich juice. The cutting is done during the hottest portion of the day, and the plant at once carried to the barn. A sharpened piece of cane is then pushed into the stalk so as to form a hook, by means of which the plant is hung on ropes stretched throughout the curing barn.

As rapidly as the leaves dry they are stripped from the stalk. The first leaves are ready to strip in about eight days, and the last in another eight. The midrib is removed from the leaf as soon as possible, and the stripped leaf made up into small rolls. These rolls are placed in boxes holding about fifty pounds, and, by means of weights on a fifteen foot lever, a pressure of about seven thousand pounds is applied. This force expresses a portion of the juice. Every day for a week or two the pressure is removed, and the leaves allowed to absorb the oxidized juice. After the first week the pressure is left on for longer and longer intervals, and at the end of three months the tobacco has become oily black in colour, and developed a rich spirituous aroma.

After being removed from the press the tobacco is made up into cylindrical rolls called "carrottes". After the leaves have been aired, a cloth twenty-four by fifteen inches is laid on a table, and

154 THE CULTURE OF TOBACCO.

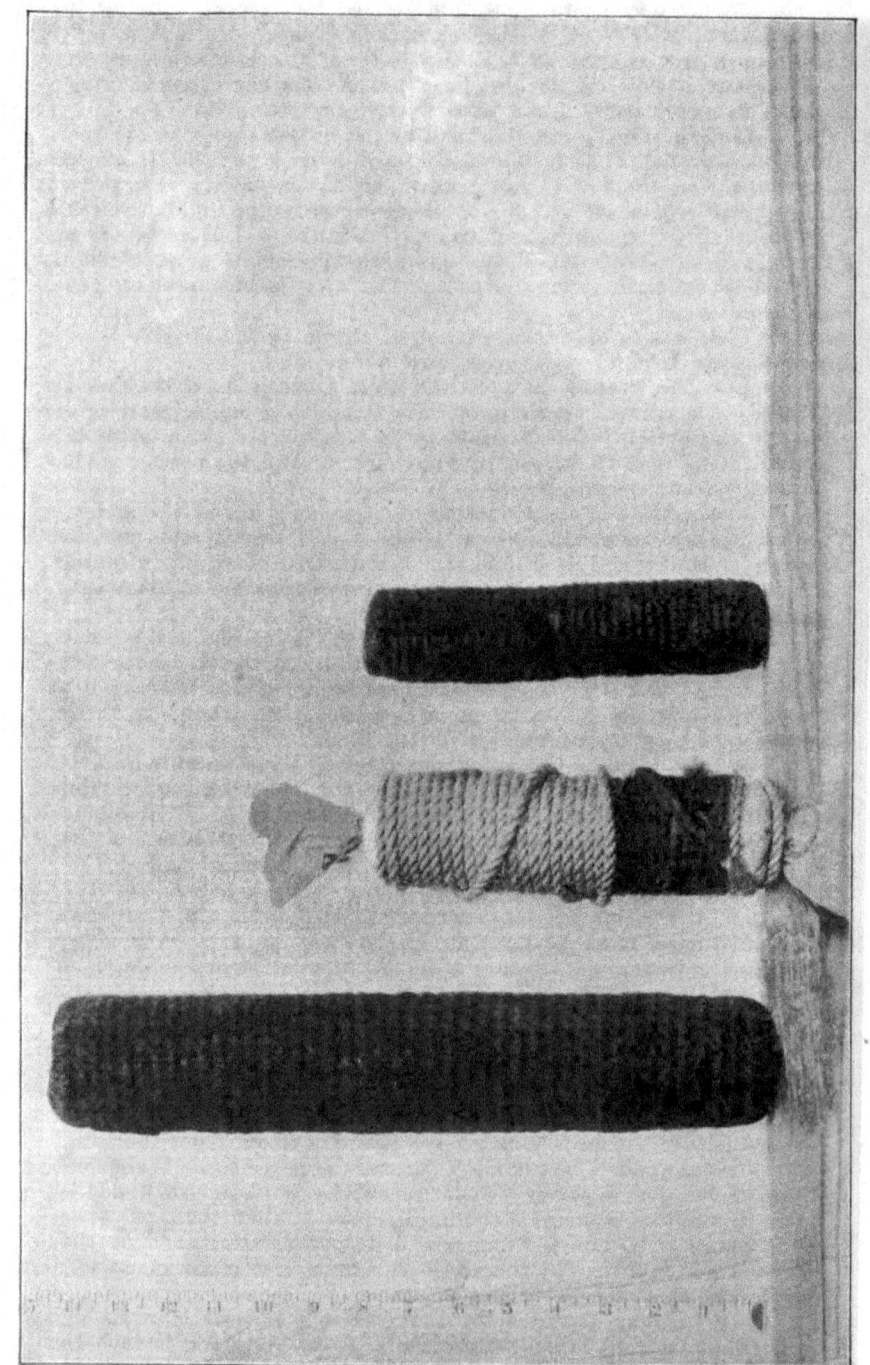

PERIQUE CARROTTES; THE ROPE AND CLOTH STILL ON THE MIDDLE ONE (U.S. DEPT. OF AG.).

on this a layer of the best wrapper leaves is laid, with their best side down, and so arranged as to have their fibres point to a longitudinal median line. A half inch of leaves is then placed on this, and after being covered with another cloth is tramped. The ends of the cloth are then doubled over for about three inches, and the whole tramped again. The tobacco is then rolled into a cylinder, and the loose ends tucked down into the hollow centre. The ends of the cloth are then tied, and the carrotte, which now weighs four pounds, and is fifteen inches long by three in diameter, is wound tightly with a one-third inch rope; the winding is done by means of a special windlass. The following day the rope is re-wound, and left on for three months, or until it is ready for the market. To market the carrottes they are packed in five hundred pound whiskey barrels. The average yield is about five hundred pounds of finished tobacco per acre. The average selling price is one shilling and ninepence per pound. Because of their peculiar system of fermentation, which the United States authorities have decided to be a process of manufacture, the Perique growers are obliged to have a manufacturer's licence, and place revenue stamps on all their tobacco.

Except by the producers themselves, Perique is seldom used in the pure form, but it is highly prized for blending with other tobaccos. A small amount is made into cigars, but the major portion is used in the best grades of pipe tobaccos in England and America. The production could be enormously increased, and no doubt would be, if others than the quiet Arcadians should take the matter up.

The aroma, like the aroma of all tobaccos, improves with age, and the memory of some nine year old Perique still haunts me.

To us Perique is important, not that we may hope to transfer its culture to Africa, but from the principle involved in its curing and fermentation. The use of pressure, and the oxidization of the juice by direct exposure to the air, may be applicable to other tobaccos, and while a Perique may not be produced, something else with a distinct pleasing odour of its own may be the result. In fact, it is known that the principle can be used with good results, for, in one of the large tobacco centres of America, there is a leaf handler who makes a business of buying up a certain grade of cheap leaf and, by the application of heat and pressure, producing an imitation Perique that is proving very acceptable to those independent manufacturers who are short of the real article.

The constant expression and absorption and the resultant oxidization of the juices appear to change a large portion of the citric and malic acids into acetic and butyric acids. The finished Perique contains only one-fourth of the citric acid, one-half or less of the nitric, and six times the acetic acid contained by the air-cured leaf. Certain volatile oils are also produced.

TOBACCO CULTURE IN SUMATRA.

Because of the unique system used, and the high quality of the leaf produced, a short description of the culture of tobacco in Sumatra may be of interest and value.

The Sumatra leaf is wonderfully thin in texture, and because of this fact, it is prized very highly for cigar wrappers. Its value is based partly on the appearance of the leaf, partly on the skill involved in the grading and packing, and partly on the economy in its use. It requires two hundred leaves to weigh a pound, and two pounds will wrap a thousand cigars.

The portion of Sumatra where this tobacco is produced is practically on the equator, and borders on the Straits of Malacca. The climate is truly a tropical one, and the temperature remains uniform throughout the year; the mean maximum for each month being about 90 degrees, the mean minimum 74 degrees, and the mean about 81 degrees. The daily temperature ranges from about 70 degrees at sunrise to 94 degrees at noonday.

The average yearly rainfall is about one hundred inches. Some rain falls every month, but the heaviest downfall is between October and December. The lowest average for any month is about three inches and the highest average eighteen inches, although months have been recorded with but a quarter of an inch, and others with thirty-five inches.

Beginning at a distance of five or ten miles back from the Straits, the tobacco belt extends for a distance of forty-five miles, and up on to the slopes of the mountains.

The soil is mostly volcanic in its origin. That on which the finest, silkiest, brown tobacco is produced is inclined to be somewhat clayey or silty. Lighter coloured tobaccos are produced on sand loams resting on a clay subsoil.

The land is leased from the Sultan for a term of seventy-five years. At least seven thousand acres are thought necessary to ensure success, for the reason that a portion of the land is permitted to lie idle each year and thus restore its fertility.

An estate is laid out by the building of a main road through it. On each side of the main road at the distance of nine hundred and sixty feet from it and from each other, smaller roads are constructed. All the land between these roads is divided into fields sixty feet wide. Each field contains about one and a third acres, and is allotted to a coolie for cultivation.

The labour is mostly done by Chinese coolies, who are secured directly from China by the planters, who club together and pay some of their number for securing them. Without the coolie, tobacco culture could not be carried on in Sumatra at all. They stand the climate well, are industrious, learn rapidly what is required of them, and are regarded as generally very satisfactory by the planters. They know absolutely nothing of tobacco culture when first secured, but must learn from those that have been at the work. Each Chinaman is given a small advance of money to live upon until the crop is harvested, and he is given another small advance at the beginning of each month. Each coolie is allotted a field by himself, and is paid by the thousand plants produced. Later he is employed at piece work in the fermentation shed and packing house.

The coolies are grouped into companies of forty, over whom is placed a foreman coolie who has been some time on the estate and understands the work. He receives a percentage based on the amount that the coolies earn. Over all the coolies and foremen is a head coolie who has been promoted to that position because of his ability. He is necessary to adjust matters between the Europeans and the

CURING BARNS IN PROCESS OF CONSTRUCTION } (FROM A DUTCH BOOK).
FERMENTATION HOUSE, SUMATRA

coolies. He gets compensation proportionate to position and services.

A coolie is given a letter of discharge when he is honourably discharged, and no planter will engage any coolie on the island who does not have such a letter. No coolie is allowed to go off the estate without a pass, and if he be found without one, he is lodged in jail until his employer is located. A reward is paid to the person who returns the coolie, and this amount is charged to the coolie's wages.

The expense of clearing the field is charged to the coolie unless he does the work himself. No more than sixteen shillings, however, is charged for any one field. After the field has been burned, and thoroughly cleaned, it is dug up to the depth of twelve inches by hand. No ploughs are used in the new fields because of the large number of roots and stumps. Ditches are dug between each two fields and empty into the main ditches at the side of the road.

Each coolie makes his own plant beds. Sometimes a large number of these beds are made so as to insure a stand of plants. The bed is protected by a shading of grass. The seed is furnished the coolie, and just enough seed for one bed is delivered at a time.

The plants are set in the field two feet apart in three foot rows, although sometimes the distances are reduced to two feet each way. Each plant is shaded by having a piece of thin light wood placed slantingly on the sunny side. These cost from thirty shillings to two pounds per ten thousand, and they are charged to the coolie's account, but he may return them and get credit when he has finished using them. Thorough cultivation by hand implements is demanded. With each cultivation the earth is banked up more and more to the plants. At the time of the second cultivation the lower leaves are broken off and buried at the base of the plant. All diseased or injured plants are destroyed, and if the plants are not too large they are replaced. The plants are topped at from fifteen to twenty-five leaves. The suckers are taken off as they appear and the worms are gathered by hand and destroyed.

When ripe, the leaves are picked off and placed in baskets, which are delivered to the curing shed. The coolie gets about twenty-six shillings per thousand for the best plants produced and as low as three shillings for the worst. If he grows a second crop from the suckers, he gets about half as much per thousand for it as for the first.

When the tobacco has been received at the curing barn from the coolie it is hung on sticks. The barns are usually kept closed at night and open in the day time. During wet weather a fire is built that does not give any smoke or odour. A watchman remains in the barn every night and regulates the ventilation according to instructions given him. From four to five weeks complete the curing.

The drying sheds are placed at the road, so that one can do for each eight fields. They are usually seventy-two feet wide, one hundred and eighty feet long, and thirty-six feet from the ground to the ridge pole. The frame is formed of material taken from the jungle. The sides are covered with rough planks, and the roof is thatched with a covering made from the leaves of a palm. The building is cheaply put up, for it is not intended to last more than two years, as the fields will then be abandoned and new ones laid out. In deserting a building the thatch is removed, rolled up, and used on

FERMENTATION PILES, SUMATRA } (FROM A DUTCH BOOK).
COOLIES GRADING TOBACCO IN SUMATRA }

the new building. The doors and windows are made so as to be used as ventilators.

When the tobacco is thoroughly cured it is roughly graded, tied into bundles or hands of about fifty leaves each, and taken in baskets to the fermentation house.

The fermentation house is substantially built of brick, and roofed with the same palm thatch as that used for the curing barns. The building is usually two hundred and forty feet long and sixty-six feet wide. In the centre of the building is a wooden platform raised three feet off the floor, and all around the platform is a free space fifteen feet wide on the floor where the coolies sit on mats and assort the tobacco. Both sides of the building are well lighted with glass windows. There is usually a small receiving room at one side of the building where the tobacco is received from the curing sheds.

The fermentation is conducted very much as is the fermentation of the better grades of cigar leaf in America. When the tobacco has been thoroughly fermented, it is carefully graded according to colour, size, texture and condition, and made into bundles of from thirty to forty leaves. These bundles are tied with a portion of the inner bark of a tree. One of the merits of Sumatra tobacco that has helped to give it a reputation is the thoroughness of the grading. A cigar maker knows when he purchases a bale of Sumatra just what there is in it, and what he can get out of it, so that he is willing to give a higher price than for a bale where the contents are unknown or doubtful.

The tobacco is placed in bales holding one hundred and seventy-six English pounds, and covered with matting. A bale is about two and-a-half feet square and a foot thick. Each bale is then distinctly marked with a description of the contents and the name of the estate, and is then ready for shipment. Nearly all of it first goes to the Amsterdam market, whence it is re-shipped to all portions of the world.

Some of the companies employ as many as sixteen thousand coolies and a large staff of experts. One company has paid an average dividend of seventy-five per cent. for many years.

COST OF PRODUCTION, PROFITS, WAGES AND YIELDS.

In round figures the American tobacco growers produce annually six hundred million pounds of tobacco, worth twelve million pounds sterling, or an average of fivepence a pound at first hand. A nice percentage of the twelve million pounds is clear profit. Whether the individual tobacco grower is making a profit or not depends largely on his own ability. Where men are employing their heads in the production of tobacco they are making money; and in many instances of exceptionally favoured location they are making money without any great use of brain power. When a locality is found where people say that they are not making money in the cultivation of tobacco and yet continue in its cultivation, something is wrong—either their methods are faulty or the conditions are not favourable. In either case, they show lack of ability, for if the methods are wrong they should discover the fact and change them, and if the locality and conditions are not favourable they should be producing some other crop than tobacco.

In a certain portion of Virginia where the farms are small, and the acreage grown on a farm is limited owing to lack of capital and executive ability on the part of the people, who are but little educated, you will be told that "ther aint no money in 'bacca round yhere." These people destroy half the value of their crop through carelessness in handling and grading. Their tobacco is often nearly ruined by mould. If due care were exercised a good living could be made out of the tobacco now lost through neglect. In this section the planter and his family do all the work, and the cost of production is unknown, as no accounts are kept. The careless methods blindly followed by them would make bankruptcy certain in any line of business.

The neighbourhood of Darlington, S. C., gives a good example of what can be done with tobacco. Up to within ten years ago, cotton was the staple crop of all the farmers. In this locality the profits of cotton raising were not very great, the soil being without much fertility, so that but little more than a mere living was made by the planters. By experiment the soil was found to be adapted to the culture of Bright tobacco. The industry was at once entered upon, and, from the beginning, the most up-to-date methods were adopted. Any features of the culture that were retained in the older tobacco districts, through stupid inertia or ancestor-worship, were discarded. During the winter of 1902-3, more than seven million pounds of Bright tobacco were sold on the Darlington market for an average price of fivepence a pound. A portion of this tobacco was produced by ignorant negro planters who placed a low grade of tobacco upon the market, so that the average received by the better planters was more than the figure named. Some received as high as £50 an acre for their crop. Careful planters say that it costs from £8 to £10 an acre to produce and market tobacco, and that they receive during a period of years at least £15 return an acre. It is easy to see in riding about this district that the farmers are prosperous. The old buildings of cotton-planting days are giving way to new ones, and the farms are being rapidly improved and developed. Land that was worth but £2 an acre several years ago is now valued at £10.

The yield of tobacco per acre is from eight to twelve hundred pounds. Where the cost of production is estimated at £8, the cost per pound of tobacco on a low yield of eight hundred pounds is two and a half pence. Low grades of leaf may sell for twopence a pound, medium at from four to sixpence, and the best grades at fifteen pence. Of the expenditure of £8, from two to three pounds is for commercial fertilizer, which is an absolute necessity on the light soil. All the work is done by negroes, who are superior as labourers to the raw African negro, although they all possess the same general traits and characteristics. A negro is paid from £1. 10s. to £2 a month in money and merchandize, and is allowed a small cabin and garden on the farm. The negro's family also works during the busy season. It is estimated that two good negroes and a mule, with extra help at planting and harvesting times, is sufficient labour for the culture of ten acres of ordinary tobacco in this section.

The White Burley tobacco of Kentucky is grown on lands that will produce splendid crops of cereals and grass, but the tobacco is regarded as the best money-making crop. The lands are valued at from £20 to £30 an acre. The wages and cost of feeding a white labourer may be put at £6 a month; the cost of production per pound at from two to three pence; and the selling price at from four to fivepence. These are averages, and there will be great variations in the different crops. An average crop may be estimated at a thousand pounds an acre of marketable tobacco, although at times two thousand are grown.

Lancaster County, Pennsylvania, is considered the wealthiest farming county in America. Tobacco is the staple crop grown and the source of most of the wealth. In this county land is valued at from £30 to £50 an acre. The yield of tobacco averages fourteen hundred pounds per acre, although two thousand pounds is not an unusual crop. The average cost of production is about four pence a pound, and the selling price about sixpence a pound. The cost of production is somewhat high when the large yields are considered, but this is owing to the great value of the land and the expensive barns used. The cost of labour may be placed at £6 a month. The tobacco produced is all cigar leaf.

In the cigar leaf section of Ohio, lands are valued at about £20 an acre. Labour costs about £6 a month. The yield of Zimmer Spanish is one thousand or twelve hundred pounds weight an acre. The average cost of production is about £10 an acre, and the selling value is from £15 to £20 an acre.

In the cigar leaf region of Connecticut, land is valued at from £40 to £60 an acre; the cost of production at £30; and the selling price at from £40 to £60 an acre. The yield is from twelve hundred to two thousand pounds an acre. This section is very wealthy and prosperous.

In Florida the lands are exceedingly cheap, where unimproved, because they are very abundant, £2 an acre being a high price for the ordinary lands. Where they have been improved, covered with canvas, and irrigated, they are valued at £200 an acre. The yield per acre of cigar tobacco may be estimated at one thousand pounds. The total cost of production up to the point where it is ready for manufacture is about twenty pence a pound, and the selling price an average of four shillings. Cuban tobacco grown in open field costs about fivepence a pound to produce, and sells at from ten to twelve pence, or even much higher.

Throughout America, the wages for a negro labourer in the tobacco field may be estimated at £2 a month, of a white labourer at £6 a month, of an expert curer on the farm £12 to £15 a month; of a cigar leaf warehouseman from £20 to £30 a month; of the manager of a tobacco-growing company from £1,000 to £2,000 a year, and of a sub-manager from £200 to £400 a year.

It is in the production of the best grades of tobacco that the most money is made. Tobacco can hardly be called a necessity of life, and men who purchase luxuries are looking for those qualities that will give them the most enjoyment. The man who will object to an extra twopence per hundredweight for flour will pay without a murmur an additional shilling for a tobacco that meets with his approval. Quality is the feature that counts.

Besides the profit that may be in tobacco culture, there is this advantage, that cured tobacco is not a very perishable article, and may be held, if necessary, for several years until the market is favourable, or may be shipped long distances with but little risk.

Tobacco is a crop that is peculiarly adapted to a new country, with cheap virgin lands and a plentiful supply of ordinary labour. The last is all important and absolutely indispensable. A constant supply of labour is necessary to the success of the industry, for if the labour supply fail during any of the critical stages of the tobacco crop, as at the topping, suckering, or harvesting times, the crop will be a failure, and the money invested be lost. The labour conditions should also be such, that the same labourers will be secured year after year, for they will greatly increase in efficiency where they continue at the same work, and learn its routine and details.

Tobacco is well adapted to culture by large companies, for there is no limit to the acreage that may be controlled by one corporation. The production of a high-grade tobacco requires the services of a high-salaried expert, as well as a corps of lesser experts. They can be secured only by a company with money and influence. The tobacco-growing companies of Florida pay most enormous returns, and one of the largest Sumatra companies pays an average dividend of seventy-five per cent., and has, out of its profits, increased its capital to about twenty times the amount paid in. This Company employs sixteen thousand workmen, as well as a corps of one hundred and sixty experts and managers.

A FEW CONCLUDING WORDS.

The development of the tobacco industry has been the work of four centuries, during which time many facts have been ascertained by experience, some learned by accident, and others by careful scientific investigation.

The planters of a country in newly taking up the culture of tobacco need not go through all the varied steps, and themselves experience all that the older countries have done, for they have at their command the accumulated knowledge of the rest of the world for four centuries. At the same time, the new country cannot start out full-fledged in the industry, for there are local, climatic and soil influences to be determined, and labour conditions to be adjusted, as well as markets to be found.

The first step in the culture is the experimental stage, and before starting this work the experimenter should be thoroughly informed as to the nature of the tobacco plant and its requirements, for if he is not, the causes of the good or bad results obtained will not be understood, and the conclusions drawn will be faulty and of little value. One year's carefully planned, intelligently observed, and accurately recorded experiments will be of more value than twenty years of experiments carelessly conducted.

Next, after favourable results have been obtained in the experimental stage, comes the period of commercial expansion. Labour must be secured, retained, and trained, because a large tobacco industry is not likely to be established with an irregular, untrained labour supply.

Then comes the question of the development of the markets. Two things will produce a market for a tobacco—merit and energy, and these should not be divorced. A superior tobacco may have difficulty in establishing itself for the reason that tobacco dealers are conservative, and look askance at an unknown product. Energy, tact and determination, however, will push an unknown leaf until its merits are familiar to all. The world really wants new superior tobaccos, but is somewhat slow in discovering them when they are produced. To secure a market for a new inferior tobacco at the present time requires "cheek" added to energy, and as soon as the element of "cheek" is removed the market is lost. A bad tobacco is not wanted unless the consumer through long usage has become accustomed to the bad tobacco and perverted his taste. The production of a bad tobacco seldom pays, and rarely merits consideration. On the other hand, the production of an unusually high-grade tobacco with an established reputation will pay under most adverse conditions of transportation and expensive labour.

Only thoroughly aged tobacco, and the best grades of the type, should be sent to the market that is being developed. Extreme care must be exercised in the packing, and the style of the package should conform with the standards and customs of the market shipped to. Honest packing should be supplemented by careful labelling of the package with the shipper's name and the description of the contents, so that credit may be given to whom it is due. Once a demand has been created, never, no matter how great the inducement may be, lower the standard of quality. The prospects of more than one tobacco have been ruined by laxity in this regard. To recover a reputation is more difficult than to create one.

Rhodesia appears to have a climate and soil in certain sections favourable to the production of high-grade tobacco It may be that certain localities are adapted to a superior leaf; if this be so, settlers in those localities are to be congratulated, for the production of high-grade cigar leaf insures prosperity. Nothing can be fully determined without experiments, which, properly conducted, will give us our answer, and point our way to the future.

THE CULTURE OF TOBACCO.

A FERTILIZER DRILL.

SPEARING TOBACCO ON THE CURING STICKS.

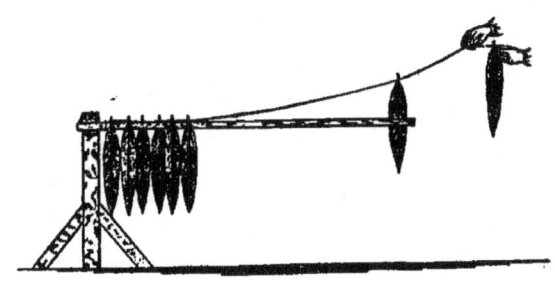

STRINGING PRIMED LEAVES.

166   *THE CULTURE OF TOBACCO.*

FIELD BASKET FOR GATHERING GREEN TOBACCO.

WAREHOUSE BASKET.

APPENDIX.

METEOROLOGICAL CONDITIONS IN THE GREAT TOBACCO REGIONS.

The following tables taken from Prof. Whitney's Bulletin on Soils (U. S. Department of Agriculture) are of interest, although their study does not, in any large way, explain the difference in the quality of the tobacco produced in the different districts. The records given are for the months during which tobacco is growing, and for this reason those for Habana are for different months than those for the other districts.

Mean monthly temperatures.

District.	Apr.	May.	June.	July.	Aug.	Sept.
	°F.	°F.	°F.	°F.	°F.	°F.
Connecticut Valley	44·80	56·50	65·90	70·20	67·70	61·11
Pennsylvania	49·90	62·00	71·80	75·90	73·30	65·63
Kentucky	57·50	65·00	75·30	77·20	75·30	69·63
Tennessee	60·60	68·10	77·00	79·50	77·93	70·53
Virginia	56·00	65·80	74·00	77·80	75·40	69·00
North Carolina	58·20	67·00	75·80	78·70	76·00	70·10
Sumatra	83·05	82·90	82·35	82·45	81·35	81·45

District.	Oct.	Nov.	Dec.	Jan.	Feb.	Mar.
Habana	78·60	75·40	72·50	72·00	73·60	75·20

Mean maximum temperatures.

District.	Apr.	May.	June.	July.	Aug.	Sept.
	°F.	°F.	°F.	°F.	°F.	°F.
Connecticut Valley	56·70	69·00	78·00	82·10	78·90	71·23
Pennsylvania	60·90	71·20	81·60	84·50	82·80	74·98
Kentucky	70·40	79·80	84·80	84·50	81·30	78·90
Tennessee	70·50	78·90	86·10	89·60	87·20	80·75
Virginia	66·60	76·20	83·60	87·20	84·20	78·20
North Carolina	69·80	78·30	86·10	87·00	84·00	78·30
Sumatra	89·35	88·80	87·50	87·80	86·90	87·15

District.	Oct.	Nov.	Dec.	Jan.	Feb.	Mar.
Habana	82·40	79·90	77·00	77·40	79·00	80·80

Mean minimum temperatures.

District.	Apr.	May.	June.	July.	Aug.	Sept.
	°F.	°F.	°F.	°F.	°F.	°F.
Connecticut Valley	35·40	46·50	55·90	60·50	58·90	50·37
Pennsylvania	39·50	49·40	59·70	64·00	61·00	54·68
Kentucky	47·90	56·20	66·20	67·60	65·60	57·35
Tennessee	51·50	59·30	67·80	71·30	68·50	62·25
Virginia	45·40	55·50	64·50	68·50	66·50	59·90
North Carolina	47·90	57·60	65·60	68·20	66·90	61·10
Sumatra	74·40	74·00	74·25	74·15	73·60	73·70

District.	Oct.	Nov.	Dec.	Jan.	Feb.	Mar.
Habana	74·10	70·70	68·00	66·90	68·00	68·40

Mean daily ranges of temperature.

District.	Apr.	May.	June.	July.	Aug.	Sept.
	°F.	°F.	°F.	°F.	°F.	°F.
Connecticut Valley	21·3	22·5	22·1	21·6	20·1	20·4
Pennsylvania	21·4	21·8	21·9	20·5	21·8	20·3
Kentucky	22·5	23·6	18·6	16·9	15·7	21·6
Tennessee	19·0	19·5	18·3	18·3	18·7	18·5
Virginia	21·2	20·7	19·1	18·7	17·7	18·3
North Carolina	21·9	20·7	20·5	18·3	17·1	17·2
Sumatra	14·9	14·8	13·2	13·6	13·3	13·4

District.	Oct.	Nov.	Dec.	Jan.	Feb.	Mar.
Habana	8·3	9·2	9·0	10·5	11·0	12·4

Records of rainfall.

District.	Averages for months.						Totals.	
	Apr.	May.	June.	July.	Aug.	Sept.	6 months.	Year.
	Inches.	Inches.	Inches.	Inches.	Inches.	Inches.	Inches.	Inches.
Connecticut Valley	3·33	3·78	3·79	4·90	4·86	3·85	24·24	49·23
Pennsylvania	3·61	4·70	3·91	4·06	3·96	3·63	23·92	43·74
Kentucky	4·98	4·46	4·25	3·97	4·12	3·06	24·86	51·70
Tennessee	5·21	3·92	4·44	3·45	3·89	3·28	24·19	52·10
Virginia	3·30	3·90	3·60	3·80	4·10	3·80	22·40	44·50
North Carolina	3·21	5·21	3·80	4·89	5·64	4·00	26·75	45·25
Sumatra	5·93	9·61	6·59	7·07	10·99	12·71	52·90	—

District.	Oct.	Nov.	Dec.	Jan.	Feb.	Mar.	6 months.	Year.
Habana	6·57	1·94	2·27	2·95	2·06	1·21	17·00	49·83

Mean relative humidity.

District.	Apr.	May.	June.	July.	Aug.	Sept.
	Per ct.	Per ct.	Per ct.	Per ct.	Per ct.	Per ct.
Connecticut Valley	63·00	66·00	70·00	70·00	74·00	74·33
Pennsylvania	70·00	75·00	76·00	76·00	73·00	82·33
Kentucky	64·00	67·00	71·00	68·00	70·00	72·00
Tennessee	61·00	64·00	69·00	70·00	70·00	70·00
Virginia	59·00	63·00	68·00	68·00	72·00	72·00
North Carolina	67·00	70·00	75·00	78·00	82·00	80·00
Sumatra	76·50	79·00	79·50	75·50	79·00	80·50

District.	Oct.	Nov.	Dec.	Jan.	Feb.	Mar.
Habana	76·90	76·10	74·60	74·40	73·60	69·30

Average number of rainy days.

District.	Apr.	May.	June.	July.	Aug.	Sept.
Connecticut Valley	8	11	9	10	9	8
Pennsylvania	10	13	10	9	9	8
Kentucky	10	12	12	8	8	7
Tennessee	11	11	11	10	8	7
Virginia	11	12	12	11	12	9
North Carolina	9	13	11	14	13	10
Sumatra	13	14	12	11	14	15

District.	Oct.	Nov.	Dec.	Jan.	Feb.	Mar.
Habana	15	10	9	8	6	6

For a determination of the adaptability of a soil for any particular crop, the mechanical analysis of a soil is of more value than a chemical analysis. By a mechanical analysis is meant a determination of the percentages of soil-grains of different sizes, and of their relation one to another; with this is also included the determination of the water-holding capacity of the soil. The following tables from Bulletin 11, United States Department of

Agriculture, are the results of the mechanical analysis of a large number of tobacco sub-soils:—

NORTHERN STATES CIGAR-LEAF SUB-SOILS.

Mechanical analyses of sub-soils.

No. of samples.	District.	Principal grade of leaf produced at the present time.	Moisture in air-dry sample.	Organic matter.	Gravel (2-1 mm.).	Coarse sand (1-0·5 mm.).	Medium sand (0·5-0·25 mm.).	Fine sand (0·25-0·1 mm.).	Very fine sand (0·1-0·05 mm.).	Silt (0·05-0·01 mm.).	Fine silt (0·01-0·005 mm.).	Clay (0·005-0·0001 mm.).
			P. ct	P. ct	P. ct	P. ct	P. ct	P. ct	P. ct	P. ct	P. ct	P. ct
9	Connecticut	Wrapper and binder.	0·76	2·53	1·03	3·26	9·92	22·62	45·47	10·41	1·36	2·32
5	Massachusetts	Do.	·61	2·20	·00	·04	·71	10·09	49·26	30·89	2·71	3·31
10	New York	Do.	1·06	2·82	1·94	2·80	9·02	24·47	32·52	15·09	3·09	7·43
5	Pennsylvania a	Do.	2·03	3·23	·67	1·23	5·87	6·62	37·18	23·41	5·21	13·80
10	Wisconsin	Binder	4·70	2·93	·59	1·09	4·98	10·34	15·68	31·04	6·01	22·76
4	Ohio	Filler	3·05	2·67	·39	·76	2·25	5·04	15·36	37·60	6·41	27·52
6	Pennsylvania b	Do.	3·61	4·47	·68	·78	·91	2·47	13·89	34·23	9·79	29·27

a River land and shaly limestone.   *b* Trenton limestone.

SOUTHERN STATES, CUBA AND SUMATRA CIGAR-LEAF SUB-SOILS.

Mechanical analyses of sub-soils.

Number of samples.	Locality.	Grade of Leaf.	Moisture in air-dry sample.	Organic matter.	Gravel (2-1 mm.).	Coarse sand (1-0·5 mm.).	Medium sand (0·5-0·25 mm.).	Fine sand (0·25-0·1 mm.).	Very fine sand (0·1-0·05 mm.).	Silt (0·05-0·01 mm.).	Fine silt (0·01-0·005 mm.).	Clay (0·005-0·0001 mm.).
			P. ct	P. ct	P. ct	P. ct	P. ct	P. ct	P. ct	P. ct	P. ct	P. ct
29	Florida, peninsula.	Main-crop wrapper, binder filler, Sucker-crop filler.	0·62	1·73	0·26	2·60	18·94	51·53	18·95	1·33	0·59	3·21
4	Florida, Gadsden County.	Same grades *a*	·58	2·68	·68	4·85	20·03	45·53	14·93	4·15	·80	5·15
4	Do.	Same grades, sub-soil.	1·18	5·69	·54	1·94	8·81	35·15	13·39	3·37	1·07	29·30
3	Texas	Same grades	·23	·46	1·63	6·58	24·55	37·05	14·16	8·90	1·59	4·70
1	California	Wrapper and filler.	1·28	3·91	2·94	5·49	19·44	27·33	12·85	13·37	2·18	10·77
8	Sumatra	Wrapper	7·48	15·41	1·41	4·39	9·95	16·15	17·17	19·11	4·35	5·00
6	Cuba (Vuelta Abajo).	Wrapper and filler.	·74	3·80	4·06	4·62	8·28	21·67	43·09	6·53	1·82	5·69
4	Cuba (Remedios).	Same, heavier	5·17	10·01	1·31	·36	·52	4·51	14·97	21·24	9·37	32·32

a A light loam, averaging 12 to 18 inches deep, overlying the red clay.

Manufacturing and Export Tobacco Sub-soils.

Mechanical analyses of sub-soils.

Number of samples.	Locality.	Description.	Moisture in air-dry sample.	Organic matter.	Gravel (2-1 mm.).	Coarse sand (1-0·5 mm.).	Medium sand (0·5-0·25 mm.).	Fine sand (0·25-0·1 mm.).	Very fine sand (0·1-0·05 mm.).	Silt (0·05-0·01 mm.).	Fine silt (0·01-0·005 mm.).	Clay (0·005-0·0001 mm.).
			P. ct	P. ct	P. ct	P. ct	P. ct	P. ct	P. ct	P. ct	P. ct	P. ct
44	Virginia and North Carolina.	Bright Yellow	1·10	2·24	2·57	6·39	13·67	22·02	23·45	14·08	5·43	8·23
55	Kentucky and Tennessee.	Export	2·23	3·00	·39	·56	·73	1·93	9·50	52·50	6·28	22·59
30	Kentucky and Ohio.	White Burley	3·48	4·42	·64	1·63	1·44	1·22	7·04	39·77	9·36	31·62
21	Virginia	Manufacturing	5·55	7·87	1·22	2·05	3·47	6·94	9·45	11·29	7·67	44·38

While the chemical analysis of soils is not always of the same value as the mechanical analysis, still it furnishes an excellent guide for the proper fertilization of the crops grown on it. In the case of the tobacco crop it may also indicate the presence of materials injurious to the combustibility of the tobacco.

CHEMICAL ANALYSES OF TOBACCO SOILS

	Moisture (100° C.)	Organic and volatile substances	Silicic Anhydride in solution	Silicic Anhydride soluble in sodium carbonate	Ferric oxide	Alumina	Manganous oxide	Lime	Magnesia	Potash	Soda	Phosphoric Anhydride	Sulphuric Anhydride	Total
NEW MILFORD, CONNECTICUT.														
Moisture at 100° C.	1·8200	—	—	—	—	—	—	—	—	—	—	—	—	1·8200
Organic and volatile substances	—	6·8600	—	—	—	—	—	—	—	—	—	—	—	6·8600
Soluble in cold hydrochloric acid	—	—	0·0535	(?)	3·1433	3·4699	0·0173	0·2839	0·7647	0·1861	0·0074	0·1984	0·0378	8·1623
Soluble in hot hydrochloric acid	—	—	0·0724	4·2426	0·3833	1·2969	0·0000	0·4404	0·0205	0·0463	0·0072	0·0218	0·0378	6·1722
Soluble in sulphuric acid	—	—	0·0000	1·0188	0·0326	2·6121	0·0000	0·1630	0·0738	0·3767	0·0027	0·0000	0·0000	4·2797
Insoluble in acids	—	—	59·9632	—	0·8540	7·4694	0·0000	1·3058	0·4302	1·3155	1·5432	0·0749	0·0000	73·0162
Total	1·8200	6·8600	65·3505		4·4132	14·8513	0·0173	1·8531	1·2892	1·9246	1·5605	0·2951	0·0756	100·3104
CLARKSVILLE, TENNESSEE.														
Moisture at 100° C.	1·3500	—	—	—	—	—	—	—	—	—	—	—	—	1·3500
Organic and volatile substances	—	4·3450	—	—	—	—	—	—	—	—	—	—	—	4·3450
Soluble in cold hydrochloric acid	—	—	0·0339	(?)	1·5800	1·4006	0·0673	0·2270	0·1218	0·0445	0·0018	0·0711	0·0105	3·5784
Soluble in hot hydrochloric acid	—	—	0·0008	3·2398	0·3167	1·8001	0·2164	0·0680	0·0714	0·0730	0·0013	0·0105	0·0000	5·8649
Soluble in sulphuric acid	—	—	0·0000	2·7239	0·1701	3·3385	0·0000	0·0547	0·0951	0·2361	0·0281	0·0000	0·0000	6·6465
Insoluble in acids	—	—	73·8330	—	0·3635	2·2123	0·0000	0·2772	0·1020	1·6484	0·2654	0·0567	0·0000	78·7585
Total	1·3500	4·3450	79·8973		2·4303	8·7515	0·3037	0·6269	0·3903	2·0220	0·2966	0·1383	0·0114	100·5633
GRANVILLE, NORTH CAROLINA.														
Moisture at 100° C.	0·6650	—	—	—	—	—	—	—	—	—	—	—	—	0·6650
Organic and volatile substances	—	1·2050	—	—	—	—	—	—	—	—	—	—	—	1·2050
Soluble in cold hydrochloric acid	—	—	0·0063	(?)	0·1775	0·3523	0·0052	0·0533	0·0098	0·0115	0·0033	0·0203	0·0090	0·6488
Soluble in hot hydrochloric acid	—	—	0·1299	0·7921	0·2115	0·4737	0·0365	0·0174	0·0128	0·0046	0·0012	0·0000	0·0050	1·6847
Soluble in sulphuric acid	—	—	0·0000	0·3818	0·1642	0·8057	0·0000	0·0365	0·0044	0·0767	0·0016	0·0000	0·0000	1·5449
Insoluble in acids	—	—	92·1631	—	0·0843	0·7748	0·0000	0·1318	0·0577	0·4117	0·2831	0·0176	0·0000	93·9541
Total	0·6650	1·2050	93·5035		0·6375	2·4065	0·0417	0·2389	0·0847	0·5045	0·2892	0·0379	0·0140	99·7025

THE CULTURE OF TOBACCO.

The following tables taken from the Tenth Census of the United States give the composition of the main types of American tobaccos:—

ANALYSES OF AMERICAN TOBACCOS.

TABLE I.—PERCENTAGE COMPOSITION OF TOBACCOS, DRIED AT 100° C.

Number of Sample.	Variety.	Nicotine.	Resin and Fatty Substances.	Starch.	Glucose.	Albuminoids (N × 6.25).	Pectic Acid (anhydride).	Citric Acid (anhydride).	Malic Acid (anhydride).	Oxalic Acid (anhydride).	Acetic Acid (anhydride).	Nitric Acid (anhydride).	Ammonia.	Cellulose (crude fibre).	Sand.	Ash, exclusive of Sand and Carbonic Acid.	Undetermined.	Total.
3	Virginia; sun-cured; for manufacturing plug tobacco	3·26	4·15	5·89	6·89	16·09	6·19	2·12	5·02	0·84	0·42	0·00	0·33	9·58	0·55	12·41	26·26	100·00
5	Virginia; fired-cured; for the German and continental trade; low grade	4·30	4·65	2·75	2·75	13·66	7·46	2·84	7·58	1·03	0·55	0·00	0·32	9·24	2·38	13·36	27·13	100·00
7	Tennessee, Clarksville; fire-cured; gummy; for the German and English markets; soil: heavy, rich loam	5·20	4·99	3·54	0·00	16·54	6·01	2·99	5·51	1·30	0·39	1·55	0·98	9·08	2·25	14·37	24·61	100·00
19	Kentucky, Mason county; air-cured; for cutting or plug tobacco	3·12	5·34	4·45	0·00	15·98	7·49	4·05	9·26	2·18	0·64	0·00	0·48	12·18	0·66	16·06	18·11	100·00
10	North Carolina, Granville county; bright wrapper; grown on white or light gray sand	2·70	5·73	6·71	16·39	8·75	5·97	0·43	7·41	0·46	0·53	0·00	0·19	9·13	1·26	8·49	25·85	100·00
28	Louisiana "Perique" tobacco; "cured in its juices"	4·32	6·28	2·45	0·00	15·80	6·66	1·18	3·94	3·49	1·62	0·00	0·76	9·08	4·17	13·30	26·95	100·00
37	Louisiana "Perique" tobacco; air-cured	4·25	7·26	2·79	0·00	16·50	7·43	4·31	7·80	2·06	0·28	1·65	1·65	8·30	0·76	15·54	19·32	100·00
35	Connecticut Seed-Leaf, New M.1-ford; soil: rich loam	4·06	4·29	3·22	0·00	18·09	6·29	5·80	10·09	0·92	0·31	3·23	0·65	10·61	1·34	15·10	16·00	100·00
30	Connecticut Seed-Leaf, Hartford; sandy soil	1·14	2·93	3·14	0·00	17·33	11·24	4·95	5·04	0·95	0·48	2·39	0·62	15·23	1·48	18·56	14·52	100·00
34	Pennsylvania Seed-Leaf, Lancaster county	1·04	4·02	3·67	0·00	14·62	12·59	1·61	5·46	0·94	0·57	0·00	0·22	15·12	1·64	17·98	20·52	100·00
16	Ohio Seed-Leaf	1·92	3·87	3·19	0·00	15·30	7·46	3·46	6·58	1·42	0·42	3·41	0·92	12·87	1·85	14·22	23·11	100·00
22	New York State Seed-Leaf	2·35	3·02	2·63	0·00	16·26	8·89	4·42	8·27	1·11	0·41	2·29	1·20	12·15	1·64	15·50	18·56	100·00
25	Wisconsin and Illinois Seed-Leaf	0·86	3·28	4·15	0·00	20·34	11·61	2·99	6·88	1·07	0·68	1·22	0·63	12·97	1·53	15·43	18·36	100·00

TABLE II.—ASH ANALYSES.

Number of Sample.	Variety.	Total Ash.	Ash, exclusive of Sand and Carbonic Anhydride.	Potash.	Soda.	Lime.	Magnesia.	Ferric Oxide.	Alumina.	Manganous Oxide.	Phosphoric Anhydride.	Sulphuric Anhydride.	Silicic Anhydride.	Chlorine.
3	Virginia; sun-cured; for manufacturing plug tobacco	14·29	12·41	34·16	0·26	31·76	7·91	0·58	1·22	0·00	3·81	4·99	1·39	13·92
5	Virginia; fire-cured; for the German and continental trade; low grade	17·42	13·36	26·55	0·22	36·96	11·51	0·95	1·81	0·00	3·23	4·27	3·20	11·21
7	Tennessee, Clarksville; fire-cured; gummy; for the German and English markets; soil: heavy, rich loam	19·23	14·37	33·15	0·15	36·48	11·85	0·51	0·95	0·25	4·42	6·16	3·42	2·06
10	Kentucky, Mason county; air-cured; for cutting or plug tobacco	21·85	16·06	39·51	0·86	39·80	5·34	1·56	0·51	0·13	6·09	4·52	1·20	0·48
10	North Carolina, Granville county; bright wrapper; grown on white or light gray sand	11·19	8·40	41·56	0·47	28·12	9·78	0·69	0·29	0·11	5·23	4·53	2·75	6·57
37	Louisiana "Perique" tobacco; air-cured (leaf deprived of midrib)	19·82	15·54	30·23	0·25	37·47	12·43	1·19	0·72	0·29	6·18	6·19	1·91	3·14
35	Connecticut Seed-Leaf. New Milford; soil: rich loam	21·08	16·30	35·08	0·01	40·38	11·33	1·47	0·74	0·13	3·20	4·08	1·39	2·19
30	Connecticut Seed-Leaf, Hartford; sandy soil	22·92	18·56	41·30	0·26	28·70	7·56	2·13	0·83	0·00	3·26	3·34	1·09	11·58
34	Pennsylvania Seed-Leaf; Lancaster county	24·74	17·98	49·60	0·36	28·55	8·18	1·39	1·05	0·00	5·72	2·61	1·05	1·49
18	Ohio Seed-Leaf	19·95	14·22	33·37	0·27	34·69	17·32	1·07	0·83	trace.	4·29	3·48	3·07	1·51
23	New York State Seed-Leaf	21·12	15·50	33·13	0·39	39·26	8·60	0·74	0·56	0·16	3·61	3·78	3·67	6·10
25	Wisconsin and Illinois Seed-Leaf	20·81	15·43	38·71	1·08	33·40	12·57	0·70	0·74	trace.	3·09	3·89	4·65	0·99

TABLE III.—PROPORTION OF MINERAL INGREDIENTS, TOTAL NITROGEN, AND POTASSIUM CARBONATE, IN 100 PARTS OF THE LEAF, DRIED AT 100° C.

Number of Sample.	Variety.	Potash.	Soda.	Lime.	Magnesia.	Ferric Oxide.	Alumina.	Manganous Oxide.	Phosphoric Anhydride.	Sulphuric Anhydride.	Silicic Anhydride.	Chlorine.	Total mineral ingredients.	Nitrogen.	Potassium, Carbonate.
3	Virginia; sun cured; for manufacturing plug tobacco	4·24	0·04	3·94	0·98	0·07	0·15	0·00	0·47	0·62	0·17	1·73	12·41	3·41	1·87
5	Virginia; fire-cured; for the German and continental trade; low grade	3·55	0·03	4·94	1·54	0·12	0·24	0·00	0·43	0·57	0·44	1·50	13·36	3·21	1·41
7	Tennessee. Clarksville; fire-cured; gummy; for the German and English markets; soil: rich, heavy loam	4·77	0·02	5·24	1·71	0·07	0·14	0·02	0·64	0·89	0·40	0·38	14·37	4·77	4·78
19	Kentucky, Mason county; air-cured; for cutting or plug tobacco	6·34	0·14	6·39	0·96	0·25	0·08	0·02	0·98	0·73	0·10	0·08	16·06	3·49	8·25
10	North Carolina, Granville county; bright wrapper; grown on white or light gray sand	3·53	0·04	2·39	0·83	0·05	0·03	0·01	0·44	0·38	0·23	0·56	8·49	2·08	4·21
37	Louisiana "Perique" tobacco; air cured (leaf deprived of midrib)	4·71	0·04	5·82	1·93	0·18	0·11	0·04	0·96	0·96	0·30	0·49	15·54	5·16	4·29
35	Connecticut Seed-Leaf, New Milford; soil: rich loam	5·30	trace.	6·10	1·71	0·22	0·11	0·02	0·98	0·62	0·21	0·33	15·10	4·97	6·08
30	Connecticut Seed-Leaf, Hartford; sandy soil	7·66	0·05	5·33	1·40	0·40	0·15	0·00	0·61	0·62	0·20	2·14	18·56	4·10	6·06
34	Pennsylvania Seed-Leaf; Lancaster county	8·92	0·06	5·13	1·47	0·25	0·19	0·00	1·03	0·47	0·19	0·27	17·98	2·70	11·91
16	Ohio Seed Leaf	4·75	0·04	4·93	2·46	0·15	0·13	trace.	0·61	0·49	0·44	0·22	14·22	4·42	5·83
22	New York State Seed Leaf	5·13	0·00	6·09	1·33	0·11	0·09	0·02	0·56	0·59	0·57	0·96	15·50	4·59	4·96
25	Wisconsin and Illinois Seed-Leaf	5·97	0·17	5·17	1·94	0·12	0·11	trace.	0·46	0·60	0·72	0·15	15·42	4·23	7·83

Percentage of Nicotine in the Principal Varieties of American Tobacco.

In the following table are given the results of determinations of nicotine on samples of the principal varieties of American tobaccos. In each case the air-dried leaves (including the midrib) were finely ground, and a careful average sample taken. The nicotine was determined on the air-dried sample by the method of Schloesing, and the percentage of moisture in a separate portion by drying at 100° C. The results are stated in percentages on the sample dried at 100° C.:—

	Percentage of Nicotine.
Virginia (heavily manured lots)	5·81
Mexican Baler (heavily manured lots)	5·60
Clarksville, Tennessee (heavily manured lots)	5·29
Virginia (French Régie)	4·81
Virginia (heavy English shipping)	4·72
North Carolina Yellow ($50)	4·58
German Saucer (Kentucky)	4·55
Perique, cured in its juices (stripped from midrib)	4·32
German low grade (Virginia)	4·30
Perique, air-cured (stripped from midrib)	4·25
West Tennessee Stemmer	4·23
German (dark)	4·14
New York (Wilson's hybrid)	4·14
Connecticut Seed-Leaf (New Milford)	4·06
French Régie, A	3·90
Pennsylvania Seed Leaf	3·88
Wisconsin Havana Seed	3·82
Connecticut Seed-Leaf (Hartford)	3·49
Pennsylvania Seed-Leaf (Lancaster county)	3·47
Virginia sun cured, for plug	3·27
Perique air-cured (whole leaf)	3·25
North Carolina Yellow ($65)	3·15
Mason county, cutting or plug	3·12
Ballard county, Kentucky, bright wrapper	2·92
Owen county, Kentucky, plug fillers	2·80
North Carolina, bright wrapper	2·69
Hart county, Kentucky, bright wrapper	2·54
New York domestic Havana	2·53
Florida Seed-Leaf	2·38
New York State Seed-Leaf	2·35
Connecticut Havana Seed	2·21
Owen county, Kentucky, cutting leaf	2·19
Ohio Seed-Leaf	1·93
Sweet-scented Wisconsin and Illinois	1·33
Connecticut Seed-Leaf	1·14
Pennsylvania Seed-Leaf	1·02
Wisconsin and Illinois Seed-Leaf	0·86
Little Dutch (Miami valley)	0·63

The following Table from BULLETIN 122, NORTH CAROLINA EXPERIMENT STATION, shows the relative combustibility of leading types of tobaccos:—

BURNING QUALITY OF TOBACCO, SHOWING GLOW IN SECONDS.

Station Numbers.	Where Grown and Variety or Type.	Glow in Seconds. Average.	Range.
468	New York; Domestic Havana Eureka	14	5—35
469	Florida; Cigar leaf wrapper	117	Burned to end of strip.
470	Kentucky; London strips	8	5—25
471	Kentucky; White Burley	7	4—18
472	Massachusetts; Wilson's Hybrid	13	5—28
473	Tennessee; German Spinning Leaf	11	7—15
474	Pennsylvania; Pennsylvania Seed Leaf	175	Burned 2½ ins. and put out.
475	Connecticut; Havana Seed Leaf	16	4—45
476	North Carolina; Fancy English Strips	5	4—6
477	Virginia; Bright Wrapping Leaf	4	2—8
478	Virginia; Austrian Wrapper	5	3—8
479	Virginia; Italian Regie	31	10—90
480	Virginia; French Regie Snuff Leaf	5	3—8
481	Virginia; Fine bright mahogany wrapper	6	4—8
496	Ohio; Ohio Spanish	24	8—50
497	Virginia; Small dark wrapper	24	14—50
582	Maryland; Colory Seconds	39	15—65
583	Ohio; White Burley	4	2—6
584	Georgia; Yellow tobacco	4	2—8
585	Alabama; Yellow tobacco	6	4—10
586	Kansas; Reddish brown	6	4—10
587	Mississippi; Reddish brown	6	5—7
588	West Virginia; White Burley	52	15—150
589	Illinois; Seed Leaf	12	4—30
590	Indiana; Light Brown	10	5—25
591	Tennessee; Spanish type	15	5—40
12	North Carolina; bright wrapper, leaf cure	5	3—7
13	North Carolina; bright wrapper, stalk ,,	5	4—7
10	North Carolina; yellow, smoker, leaf ,,	8	4—13
6	North Carolina; yellow, smoker stalk ,,	7	4—12
100	North Carolina; yellow, cutter, leaf ,,	8	4—15

Absorptive Capacities of Certain Varieties of American Tobacco.

From the "Tenth Census," U.S.A.

The capacity of leaf tobacco to absorb and retain different flavouring substances added in the form of "sauces" is a matter of great importance to the manufacturer, and especially to the foreign importer of American tobaccos. In the following table are given the coefficients of absorption of some of the principal varieties used for the manufacture of chewing tobacco. These coefficients give the amount of water which each type will absorb and retain without dripping, expressed in multiples of the weight of the air-dried leaf; they do not, of course, represent the actual amount of water that a given sample will absorb and retain when subjected to the usual operations of manufacture. It may be safely assumed, however, that the results obtained in practice will stand to each other in a relation that will not vary greatly from that indicated by the theoretical coefficients of absorption, and the latter may, therefore, serve as a sufficient basis for classification and comparison.

The coefficients of absorption were determined as follows: The air-dried leaf was carefully weighed, moistened with water until it had become pliable, then loosely coiled on the bottom of a beaker and water enough added to completely cover it. The whole was then left at rest for 48 hours. The leaf was then taken out, suspended over the beaker until it had ceased to drip, and weighed. The liquid in the beaker was then evaporated to dryness on the water-bath, the residual extract dried at 100 C°., and weighed. The coefficient of absorption was determined from these data by the equation

$$\frac{a+b-c}{c}=x$$

wherein a is the weight of the wet leaf, b the weight of the dry extract, c the weight of the dry leaf, and x the coefficient of absorption. The results were as follows, viz.:

	Coefficient of absorption.
German, low grade (Virginia)	2·88
North Carolina bright wrapper	2·77
North Carolina Yellow ($65)	2·65
Owen county, Kentucky, cutting leaf	2·60
Owen county, Kentucky, plug fillers	2·55
Hart county, Kentucky, bright wrapper	2·54
North Carolina Yellow ($50)	2·39
Ballard county, Kentucky, bright wrapper	2·27
Mason county, Kentucky, cutting or plug	2·21
Régie Virginia Shipper	2·14
German Saucer	2·07
Mexican Baler	2·04
Virginia sun-cured for plug	2·02
English Shipper (Virginia)	1·95
West Tennessee Stemmer	1·92
Virginia (heavily manured)	1·92
Florida Seed-Leaf	1·79
Perique, air-cured	1·74
Ohio Seed-Leaf	1·73
Sweet-scented Wisconsin and Illinois	1·67
Clarksville, Tennessee, German	1·48
Virginia French Régie, A	1·41
Virginia German Shipper	1·12

DESCRIPTION OF TOBACCO GROWN IN DIFFERENT PARTS OF THE UNITED STATES.

These Tables, compiled from data secured by the North Carolina Experiment Station and published in Bulletin 122, are of great interest. The variety planted, the nature and value of the soil, the fertilizers applied, the purpose for which the tobacco is to be used, and the selling price are all given. The value in English pence may be approximately obtained by dividing the values in cents by two, and the values in pounds may be reached by dividing the dollars by five. The nature of the soil is described in general and in exact terms, and this fact should be remembered in drawing conclusions from the tables.

180   THE CULTURE OF TOBACCO.

DESCRIPTION OF TOBACCO GROWN

Where Grown.	Variety.	For what trade.	In what form manufactured.	Average value per pound.
Connecticut, Hartford Co.	Havana seed leaf.	Domestic	Cigar wrappers	50 cents
Connecticut, Hartford Co.	Havana seed leaf.	Cigar	Cigar wrappers	50 cents
Connecticut, Hartford Co.	Havana seed leaf.	Cigar	Exclusively for cigars.	25 cents
Connecticut, Bancroft Co.	Havana seed leaf.	Domestic	Cigar wrappers	50 cents
Florida, Columbia Co.	Cigar leaf wrapper.	Sold in Cincinnati, New York and New England.	Wrappers and fillers.	Wrappers $2 to $ fillers less.
Kentucky, Davess Co.	London strips	English	"Shag"	9 pence in London
Kentucky, Davess Co.	English leaf	English	Cut into form similar to fine cut.	8 pence in London
Kentucky, Bracken Co.	Kentucky white Burley.	Domestic and foreign.	Plug, fine cut and smoking.	10 cents
Kentucky, Graves Co.	German shipping tobacco.	Bremen	Spinning and cigar wrappers.	8½ cents
Massachusetts, Hampshire Co.	Wilson's Hybrid Havana.	Cigar	Cigar wrappers	15 to 25 cents
Massachusetts, Hampshire Co.	Havana seed leaf.	...	...	20 cents
New York, Onondaga Co.	Wilson's Hybrid Havana.	Cigar	Wrappers, binders and fillers.	Wrappers 30 cent binders 12 cent fillers 7 cents.
New York, Onondaga Co.	Domestic Havana Eureka.	Cigar	Wrappers, binders and fillers.	Wrappers 30 cent binders 12 cent fillers 7 cents.
North Carolina, Alamance Co.	Fancy English strips.	English manufacture.	Plug	45 cents, raw values
North Carolina, Wake Co.	Lemon wrapper	American	Wrapping fine grades of plug.	50 to 75 cents
North Carolina, Wake Co.	Mahogany wrapper.	American	Wrapping fine grades of plug.	50 to 75 cents
North Carolina, Wake Co.	Cutter	American	Cigarettes	27 to 30 cents
North Carolina, Wake Co.	Fine Smoker	American	Granulated smoking.	15 cents
Ohio, Miami Co.	Ohio seed leaf	Cigar	Binders' fillers and wrappers.	8 cents
Ohio, Miami Co.	Little Dutch	Cigar	Cigar fillers	13 cents
Ohio, Miami Co.	Zimmer Spanish	Cigar	Fillers and binders	11 cents
Ohio, Highland Co.	Large white Burley.	Cincinnati and Louisville.	Fine cut chewing	18 to 20 cents
Ohio, Highland Co.	Good red leaf of white Burley.	Cincinnati and Louisville.	Plug	18 to 20 cents
Ohio Highland Co.	Good bright leaf	Cincinnati and Louisville.	Plug and fine cut	20 to 25 cents

DIFFERENT PARTS OF THE UNITED STATES.

Average yield per acre.	Character of soil.	Kind and amount of fertilizers applied per acre.	Does Land depreciate with continued cropping?	Average value of Tobacco Lands per acre.
1,800 lbs.	Sandy loam	Cotton seed meal 2,000 lbs. and cotton hull ashes 1,200 lbs.	No	$150 to $200.
1,800 lbs.	Light loam	600 lbs. sul. potash, 1 ton cotton seed meal, 400 lbs. ground bone and 300 lbs. lime broadcast.	No, it improves	$200.
1,800 lbs.	Sandy loam	1,500 lbs. cotton seed meal, 500 lbs. castor pomace, 1,000 lbs. cotton hull ashes, 800 lbs. bone phos.	No, with proper fertilization.	$100.
1,800 lbs.	Sandy loam	Cotton seed meal 2,000 lbs. and cotton hull ashes 1,200 lbs.	No	$150 to $200.
500 to 1,000 lbs.	Light sandy loam	On new hammock none, on old land 200 to 2,000 lbs. mixed fertilizer.	No, if rotation of crops is practised.	$3 to $30.
1,000 lbs.	Clay loam and sandy loam	None	Yes, after three years from clearing.	$25 to $100.
1,000 lbs.	Clay loam and sandy loam.	None	Yes, after three years from clearing.	$25 to $100.
1,000 lbs.	Limestone	None	Yes	$50.
800 lbs.	Clayey loam	None	Yes	$15.
1,400 to 1,800 lbs.	Light sandy loam.	Sul. potash 600 lbs., dry fish 1,000 lbs., cotton seed meal 1,000 lbs., and barnyard manure 6 cords.	No, it improves	$200.
1,600 lbs.	Sandy loam	Cotton seed meal 2,000 lbs. and sul. potash 500 lbs.	No	$200.
1,000 to 1,200 lbs.	Sandy, gravelly loam.	Barnyard manure, ashes, cotton seed meal, plaster, etc., say 500 lbs. commercial fertilizer.	No, with proper fertilization.	$50 to $250.
Average 1,100 lbs.	Sandy and gravelly loam.	Barnyard manure, ashes, cotton seed meal, plaster, etc., say 500 lbs. commercial fertilizer.	No, with proper fertilization.	$50 to $250.
100 lbs. of this grade with 400 lbs. poorer grade	Sandy subsoil	400 lbs. ammoniated fertilizer	Yes	$10.
500 to 600 lbs.	Light sandy	...	...	...
600 to 700 lbs.	...	...	...	...
500 pounds	Light gray with yellow subsoil.	400 to 500 lbs. ammoniated fertilizer for tobacco	Yes	$15 to $40.
...	...	...	...	...
1,200 pounds	Sugar tree soil, clay.	Manure	Yes	$75.
About 800 lbs.	Sugar tree lands	Manure	Yes, not over two crops in succession.	$75.
900 pounds	Bottom lands	Manure	Yes	$75.
1,800 to 2,000 lbs., including all grades	Rich limestone soil.	This was grown in old house lot.	No, if well manured yearly.	$50 to $75.
1,800 to 2,000 lbs., including all grades	Rich limestone soil.	This was grown in old house lot.	No, if well manured yearly.	$50 to $75.
1,800 to 2,000 lbs., including all grades	Rich limestone soil.	This was grown in old house lot.	No, if well manured yearly.	$50 to $75.

DESCRIPTION OF TOBACCO GROWN IN DIFFEREN[T]

Where Grown.	Variety.	For what trade.	In what form manufactured.	Average value per pound.
Ohio, Montgomery Co.	Little Dutch filler	Cigar	Cigar fillers	8 to 9 cents
Ohio, Montgomery Co.	Ohio seed leaf wrapper.	Cigar	Cigars	About 6 cents to raise
Ohio, Montgomery Co.	Zimmer Spanish filler.	Cigar	Cigar fillers	About 9 cents to raise
Ohio, Belmont Co.	White Burley air-cured.	Home consumption	Chewing	7 cents
Ohio, Belmont Co.	White Burley fire-cured.	European		10 cents
Tennessee, Montgomery Co.	French B	France	Cigars and smoking tobacco.	6¼ to 7 cents
Tennessee, Montgomery Co.	Italian B	Italy	Cigars	7 to 7½ cents
Tennessee, Montgomery Co.	Austrian B	Austria	Cigars	9 to 11 cents
Tennessee, Montgomery Co.	Stemming leaf for English strips.	Great Britain	Twist for chewing and cut for pipe.	5¼ to 7 cents
Tennessee, Montgomery Co.	European cigar wrapper.	Germany, Switzerland and Austria.	Cigars	10 to 13 cents
Tennessee, Montgomery Co.	German filler	Germany, Norway and Sweden.	Twist for chewing	6 to 6½ cents
Tennessee, Montgomery Co.	German spinning leaf.	Germany, Norway and Sweden.	Twist for chewing	7½ to 11 cents
Tennessee, Montgomery Co.	Austrian wrapper.	Austria	Cigar wrapper	11 cents
Tennessee, Montgomery Co.	Dark wrapper	United States	Plug	12 cents
Tennessee, Montgomery Co.	Italian style	Italian	Cigars and snuff	7½ cents
Tennessee, Montgomery Co.	German spinner.	German	Twist for chewing	9 cents
Tennessee, Montgomery Co.	French style	France	Cigars and cigarettes	6½ cents
Virginia, Henrico Co.	French Régie snuff leaf	French Régie	Snuff	7 cents
Virginia, Henrico Co.	Italian Régie cigar wrapper	Italian Régie	Cigars	14 cents
Virginia, Henrico Co.	Austrian Régie cigar wrapper	Austrian Régie	Cigars	14 to 16 cents
Virginia, Henrico Co.	Italian Régie cigar filler	Italian Régie	Cigars	10 cents
Virginia, Henrico Co.	Austrian Régie cigar filler	Austrian Régie	Cigars	13¼ cents
Virginia, Henrico Co.	Bremen Saucers	Chiefly Norway	Twist and smoking	12 cents
Virginia, Dinwiddie Co.	Large dark wrapper	Domestic and export	Plug	18 cents

PARTS OF THE UNITED STATES—*continued.*

Average yield per acre.	Character of soil.	Kind and amount of fertilizers applied per acre.	Does Land depreciate with continued cropping?	Average value of Tobacco Lands. per Acre.
900 pounds	Red clay is best adapted and produces best quality and flavour	Clover sod and stable manure preferred.	Yes, unless liberally manured with stable manure each year.	$100.
1,200 lbs.	Black sandy loam preferred	Clover sod, with 4 to 8 tons stable manure preferred.	Yes, unless liberally manured with stable manure each year.	$100.
800 lbs.	Red clay is best suited, and produces best quality.	Stable manure and other fertilizers are used in a very limited way.	Yes, without the use of fertilizers.	$100.
1,000 lbs.	Dark sandy loam.	... ...	Yes	$40.
1,000 lbs.	Sandy loam, new ground.	None on new land	Yes	$40.
800 to 900 lbs.	Thinned clay limestone foundation.	Stable manure according to supply and need.	Yes, except by rotation of crops.	$10 to $20.
900 to 1,200 lbs.	Red clay, with limestone foundation.	Stable manure and commercial fertilizer.	Yes, except by rotation of crops.	$20 to $30.
900 to 1,100 lbs.	Red clay, with limestone foundation.	Stable manure according to supply and need.	Yes, except by rotation of crops.	$20 to $30.
1,000 to 1,300 lbs.	Red clay, with limestone foundation.	Stable manure, with about 200 lbs. commercial fertilizer.	Yes, except by rotation of crops.	$15 to $30.
1,000 to 1,200 lbs.	Deep red clay, with limestone foundation.	Stable manure, with about 200 lbs. commercial fertilizer.	Yes, except by rotation of crops.	$20 to $40.
1,000 to 1,200 lbs.	Red clay, with limestone foundation.	Stable manure according to supply and need.	Yes, except by rotation of crops.	... ...
1,000 to 1,200 lbs.	Red clay, with limestone foundation.	Stable manure according to supply and need.	Yes, except by rotation of crops.	$25 to $40.
1,000 lbs.	Limestone	Tobacco grower, about 75 lbs. per acre.	Yes	$30 to $40.
1,000 lbs.	Limestone	Tobacco grower, about 50 to 75 lbs.	Yes	$30 to $40.
800 lbs.	Limestone	Tobacco grower, 50 to 75 lbs. per acre.	Yes	$30 to $40.
900 lbs.	Limestone	Tobacco grower, 50 to 75 lbs. per acre.	Yes	$30 to $40.
700 lbs.	Limestone	Tobacco grower, 50 to 75 lbs. per acre.	Yes	$30 to $40.
550 lbs.	Clay	600 lbs. sundries	Yes	$5.
800 to 1,000 lbs.	Clay	600 lbs. sundries	Yes	$5.
800 to 1,000 lbs.	Clay	600 lbs. sundries	Yes	$5.
800 to 1,000 lbs.	Clay	600 lbs. sundries	Yes	$5.
800 to 1,000 lbs.	Clay	600 lbs. sundries	Yes	$5.
500 to 600 lbs.	Sandy and heavy.	600 lbs. sundries	Yes	$5.
1,200 to 1,500 lbs., whole crop.	Red or gray, with clay subsoil.	Stable manure, with some commercial fertilizer.	Not with intelligent farming.	$8 to $10.

DESCRIPTION OF TOBACCO GROWN IN DIFFEREN

Where grown.	Variety.	For what trade.	In what form manufactured.	Average value per pound.
Virginia, Dinwiddie Co.	Small dark wrapper	Principally export	Plug and twist	11 to 12¼ cents
Virginia, Dinwiddie Co.	Austrian wrapper	Austria, Italy and other countries where dark tobaccos are used	Plug and cigars	15 to 18 cents
Virginia, Dinwiddie Co.	Fine Australian filler.	Australia and other countries.	Plug	8 to 10 cents
Virginia, Pittsylvania Co.	Fine dark Virginia filler.	... ...	Fillers for plug and twist.	12¼ cents
Virginia, Pittsylvania Co.	Fine bright Virginia filler.	... ...	Plug, twist, fine cut and plug cut.	About 12 cents
Virginia, Pittsylvania Co.	Fine bright Virginia wrapper.	... ...	Wrapping fine bright plug and twist.	45 to 65 cents
Virginia, Pittsylvania Co.	Fine bright mahogany wrapper.	... ...	Wrapping fine bright plug and twist.	25 to 40 cents
Virginia, Pittsylvania Co.	Bright export	English	Cigarette and smoking.	15 cents
Virginia, Pittsylvania Co.	Export bright strips.	English	Chewing and smoking.	18 cents
Virginia, Pittsylvania Co.	Export bright cutters.	England and other countries.	Cigarette and smoking.	16 cents
Virginia, Charlotte Co.	Cutting lug	Canadian	Smoking	7 cents
Virginia, Pittsylvania Co.	Bright wrapping leaf.	English	Smoking	25 cents
Virginia, Charlotte Co.	Common filler	Domestic	Chewing	6 cents
Virginia, Pittsylvania Co.	Canadian filler	Canadian	Smoking plug	10 cents
Virginia, Pittsylvania Co.	Fine cigarette cutter.	English	Cigarettes	24 cents in strips; cents in leaf.
Virginia, Pittsylvania Co.	Wrapper strips	English	Smoking	30 cents in strips; cents in leaf.
Virginia, Pittsylvania Co.	Medium bright wrapper.	Domestic	Plug	20 cents
Virginia, Pittsylvania Co.	English strip	English	Cigarettes	16¾ cents in strips; cents in leaf.
Virginia, Pittsylvania Co.	Fine bright wrapper.	Domestic	Plug	30 cents
Virginia, Pittsylvania Co.	Cutting lug	Domestic	Cigarettes	12 cents
Virginia, Pittsylvania Co.	English strip	English	Smoking	9 cents in strips; cents in leaf.
Virginia, Pittsylvania Co.	English strip	English	Cigarettes	16½ cents in strips; cents in leaf.
Pennsylvania	Pennsylvania seed-leaf.	Cigar	Cigar fillers and wrappers.	Wrappers 20 cents fillers, 7 cents.

PARTS OF THE UNITED STATES—*continued*.

Average yield per acre.	Character of soil.	Kind and amount of fertilizers applied per acre.	Does Land depreciate with continued cropping?	Average value of Tobacco Lands per acre.
1,000 lbs., whole crop.	Red or gray, with clay subsoil.	Stable manure, with some commercial fertilizer.	Not with intelligent farming.	$8 to $10.
1,200 to 1,500 lbs., whole crop.	Red or gray with clay subsoil.	Stable manure with some commercial fertiliser.	Not with intelligent farming.	$8 to $10.
1,000 lbs. whole crop.	Red or gray with clay subsoil.	Stable manure, with some commercial fertilizer.	Not with intelligent farming.	$8 to $10.
...	...	...	...	...
...	...	...	...	...
...	...	...	...	...
...	...	...	...	...
500 lbs.	Light gray	200 lbs. ammoniated fertilizer	Yes	$8 to $10
600 to 700 lbs.	Gray	200 to 400 lbs.	Yes, but may be reclaimed by sowing in grass.	$12.
700 lbs.	Light gray or sandy	200 to 400 lbs.	Not when properly cared for.	$12.
400 lbs.	Light gray	Ammoniated fertilizer	Yes	$10.
500 lbs.	Light gray	Ammoniated fertilizer	Yes	$10.
500 lbs.	Light gray	Ammoniated fertilizer	Yes	$10.
500 lbs.	Light gray	Ammoniated fertilizer	Yes	$10.
400 lbs.	Light gray	Ammoniated fertilizer	Yes	$10.
500 lbs.	Light gray	Ammoniated fertilizer	Yes	$10.
500 lbs.	Light gray	Ammoniated fertilizer	Yes	$10.
400 lbs.	Light gray	Ammoniated fertilizer	Yes	$10.
500 lbs.	Light gray	Ammoniated fertilizer	Yes	$10.
400 lbs.	Light gray	Ammoniated fertilizer	Yes	$10.
500 lbs.	Light gray	Ammoniated fertilizer	Yes	$10.
500 lbs.	Light gray	Ammoniated fertilizer	Yes	$10.
About 1,300 lbs.	Generally limestone; also some red sandstone and gravel.	Principally barnyard manure and sometimes commercial fertilizers.	No depreciation known.	$175 to $250.

LONDON
WATERLOW AND SONS LIMITED, PRINTERS,
LONDON WALL.